W9-ARZ-132

CLOUD PHYSICS AND CLOUD SEEDING

LOUIS J. BATTAN was born in 1923 in New York City. After having entered the City College of New York, he enlisted in the U. S. Army Air Force in World War II and later became a weather officer. With a group of selected Air Force lieutenants, he attended radar and electronics classes at Harvard and M.I.T. and was trained in the use of radar equipment for making special meteorological observations. Following the war, he returned to his studies. In 1946, he received his B.S. degree from New York University and joined the U. S. Weather Bureau, where he worked for three years on the Thunderstorm Project, a research program. Graduate work at The University of Chicago resulted in an M.S. degree in 1949 and a Ph.D. in 1953. He remained there after graduation and continued his work on the physics of clouds and precipitation.

In 1958, Dr. Battan joined the staff of The University of Arizona in Tucson as a Professor of Meteorology and the Associate Director of the Institute of Atmospheric Physics. Since then he has continued to do research in the various areas of physical meteorology, including the effects of cloud seeding. Professor Battan was a member of the Council of the American Meteorological Society in 1959–61, serving on its Committee on Radar Meteorology and Committee on Severe Storm and Cloud Physics, and is an associate editor of the *Journal of the Atmospheric Sciences,* formerly the *Journal of Meteorology*. He received the A.M.S. Meisinger Award for 1962, given in recognition of his research achievements. Professor Battan is the author of *Radar Meteorology* (University of Chicago Press, 1959), the first textbook on the subject, and of many articles for scientific journals. This is his third book for the Science Study Series, the previous ones being *The Nature of Violent Storms,* published in 1961, and *Radar Observes the Weather,* published in 1962.

CLOUD PHYSICS AND CLOUD SEEDING

by Louis J. Battan

GREENWOOD PRESS, PUBLISHERS
WESTPORT, CONNECTICUT

Library of Congress Cataloging in Publication Data

Battan, Louis J
Cloud physics and cloud seeding.

Reprint of the ed. published by Anchor Books, Garden City, N. Y., which was issued as no. S29 of the Science study series.
Bibliography: p.
Includes index.
1. Cloud physics. 2. Rain-making. I. Title.
[QC921.5.B3 1979] 551.5'77 78-25711
ISBN 0-313-20770-4

First published in 1962 by Doubleday & Company, Inc., New York

Reprinted with the permission of Doubleday & Company Inc.

Reprinted in 1979 by Greenwood Press, Inc.
51 Riverside Avenue, Westport, CT 06880

Printed in the United States of America

10 9 8 7 6 5 4 3 2 1

THE SCIENCE STUDY SERIES

The Science Study Series offers to students and to the general public the writing of distinguished authors on the most stirring and fundamental topics of science, from the smallest known particles to the whole universe. Some of the books tell of the role of science in the world of man, his technology and civilization. Others are biographical in nature, telling the fascinating stories of the great discoverers and their discoveries. All the authors have been selected both for expertness in the fields they discuss and for ability to communicate their special knowledge and their own views in an interesting way. The primary purpose of these books is to provide a survey within the grasp of the young student or the layman. Many of the books, it is hoped, will encourage the reader to make his own investigations of natural phenomena.

The Series, which now offers topics in all the sciences and their applications, had its beginning in a project to revise the secondary schools' physics curriculum. At the Massachusetts Institute of Technology during 1956 a group of physicists, high school teachers, journalists, apparatus designers, film producers, and other specialists organized the Physical Science Study Committee, now operating as a part of Educational Services Incorporated, Watertown, Massachusetts. They pooled their knowledge and experience toward the design and creation of aids to the learning of physics. Initially their effort was supported by the National

Science Foundation, which has continued to aid the program. The Ford Foundation, the Fund for the Advancement of Education, and the Alfred P. Sloan Foundation have also given support. The Committee has created a textbook, an extensive film series, a laboratory guide, especially designed apparatus, and a teachers' source book.

PREFACE

Before 1946, few meteorologists were interested in changing the weather. Most of them were concerned with observing, explaining, and forecasting various types of conditions. Not until Vincent J. Schaefer and Irving Langmuir conducted some simple, but dramatic, experiments did scientists begin to realize that it was possible to modify certain types of cloud systems.

Over the last fifteen years we have been trying to develop schemes not only to modify, but also to control the weather. To date the progress has been somewhat disappointing. But in the course of carrying out weather-modification studies, we have learned a great deal about the natural processes of cloud and rain formation.

The realization that it was within the realm of possibility to increase rainfall, suppress damaging hail, and alter the course of severe storms stimulated basic research activity all over the world. As a result, our knowledge of atmospheric phenomena is much greater today than it was only a decade ago. There still are many aspects of the problem that are shrouded with various degrees of uncertainty. However, as continued research pierces through the fog, and we learn more secrets of nature, it seems inevitable that we will learn how to modify and control the weather.

The number of atmospheric scientists working on this problem is distressingly small in view of its tremendous importance. More vigorous and imaginative people are needed. History has shown that some of the greatest scientific discoveries have been made by young men and women who have had the daring to think beyond the accepted boundaries of existing knowledge. The atmospheric sciences need this kind of person. The challenge of controlling the weather is one we must meet and conquer.

My thanks are extended to Mr. John H. Durston for his editorial assistance and to Mr. Morgan Monroe for his encouragement.

Louis J. Battan

CONTENTS

CLOUD PHYSICS AND CLOUD SEEDING

Chapter 1

INTRODUCTION

Is there life as we know it on other planets or their satellites? Nobody can be sure. But with so many millions of planets in the universe it might some day be discovered that there are other earths. At this time, however, we can be certain that on some of the heavenly bodies in our own solar system there cannot be people like the man next door, or roses like those in the back yard. The moon is such a place. You might ask, "How can you be so sure?"

Some day when the sky is clear and you can get close to any kind of telescope, take a good look at the moon. If a telescope is not handy, go to the library and get an astronomy book showing photographs of the surface of the moon. You will notice that many of its features can be seen clearly. Sometimes very small details of the moon's topography can be observed.

Now, what do you think you would see if you were standing on the moon looking at the earth? Instead of clearly seeing the surface you would observe clouds of various sizes and shapes. The surface of the earth in many places would appear fuzzy because of the haze. The reason for these things is fairly obvious. On the planet earth there is a plentiful supply of that very important substance—water.

On the moon there is no water, hence no water vapor in its atmosphere and consequently no clouds. Lacking water, no life as we know it can exist.

When we say there is a great deal of water on the earth, we refer to all kinds of water—fresh, salty, and very salty. At one time very few people worried about the supply of fresh water. The population of the world was small, most large cities were situated on the shores of lakes or well-fed rivers, industrial demands for water were not excessive, and irrigation was not widely practiced. But with the Industrial Revolution of more than a hundred years ago, the Agricultural Revolution of the last twenty or thirty years, and the great growth of population that has been going on for a long time, the demands for fresh water have soared. All over the world scientists and politicians have begun to realize that without water there can be no growth—of people, plants, or societies.

For many decades intensive research on better utilization of existing water supplies has been going on. Serious attempts are now being made to develop economical means of desalinating sea water and brackish inland waters. During the last fifteen years there also have been many studies aimed at the development of techniques for squeezing more water out of the atmosphere. These efforts for the most part have not been fruitful. Later chapters will deal with cloud-modification research. At present let us recognize that one of the chief reasons for the failure to stimulate rainfall is that we do not know in detail how nature makes rain or snow. In this short book we shall discuss certain aspects of the physics of the formation of clouds and precipitation. This type of information forms the foundation on which cloud physicists plan experiments to fill in the gaps in our knowledge. It serves as the basis for the design of cloud-modification tests. But before we embark on a discussion of the clouds themselves, it is of some value to get a better idea of the water picture on the planet earth.*

* Much of the information contained here has come from an article by J. E. McDonald in *Advances in Geophysics,* Academic Press, Vol. 5, 1958, pp. 223–98.

It has been estimated that the total quantity of water in any form is equal to about 10^{18} tons. For those encountering this notation for the first time, this is equal to the numeral 1 followed by eighteen zeros, that is, 1,000,000,000,000,000,000 tons. Only about 0.01 per cent of this water is in rivers and lakes. As much as 5 per cent is frozen in glaciers and stored underground between rocks and sand. The quantity of water vapor in the atmosphere has been estimated to be 0.001 per cent of the total stock of terrestrial water. Table 1 gives a summary of the relative amounts.

TABLE 1. QUANTITY OF WATER IN VARIOUS FORMS

	Tons	*Per cent of total*
Oceans	9.5×10^{17}	94.9
Glaciers and ground water	5×10^{16}	5.
Rivers and lakes	10^{14}	0.01
Atmospheric vapor	10^{13}	0.001

Clearly, the fraction of water in the form of atmospheric vapor is small relative to the other sources. Nevertheless, the total quantity is still quite substantial. If we take the population of the world to be about 2.7 billion, the atmospheric vapor, if equally divided, would entitle each person to 3700 tons of water. Further, it has been estimated that the entire supply of water vapor is replenished on an average of once every twelve days. Thus, if all the water vapor could be made to fall as rain, which could be collected and distributed, each person would have about 310 tons of water per day. This quantity amounts to a substantial 75,000 gallons.

Of course, such calculations merely represent an interesting game of numbers. We must immediately recognize that only the rain or snow which falls on land areas can be of any immediate use.

Scientists have estimated that the rainfall over the earth averages about 40 inches per year. This yields a total of about 8×10^{11} tons per day. Since the continents cover about one quarter of the earth's surface, the quantity of precipitation that can possibly be used is about 2.5×10^{11} tons per day. If this quantity could be equally divided, it would amount to about 8 tons (or more than 20,000 gallons per day) for every human being.

It is of interest to note that the average quantity of water used per person in the United States is about 1650 gallons per day. Of this amount 90 per cent is divided equally between industry and agriculture, while the remaining 10 per cent is used by householders.* It would appear from these figures that there is enough fresh water to supply all needs. Unfortunately, this is not the way things work out. As so often happens, the averages mentioned are misleading. In certain parts of the world there is much too much rain, and in others much too little. Furthermore, it does not all fall at the times when it is most needed.

For these reasons water is pumped from nature's reservoirs—lakes, rivers, and subterranean storage regions. Even this procedure is not sufficient. Oceans are an important potential source of fresh water, but so far the costs of taking out the salts and other minerals are too high. By means of the best practical techniques (in 1962), we would still have to pay one dollar per 1000 gallons of fresh water. This price is about three times too high. There are hopes that concentrated research efforts in various parts of the world will succeed in bringing down the cost.

Meteorologists have concerned themselves with the development of means for increasing rainfall. Although techniques presently available are not at all likely to lead to a significant change in the water budget of the earth as a whole, there are indications that it may be possible to produce small but important changes in the

* From B. F. Dodge, *American Scientist*, Vol. 48, 1960, p. 477.

clouds and precipitation over a limited region. To understand how this may be done, it is necessary first to examine the physical mechanisms of the formation of clouds and precipitation in nature.

Chapter 2

CONDENSATION NUCLEI—THE BUILDING BLOCKS OF CLOUDS

One of the joys of living by the sea is the pleasure of watching the waves splashing onto the shore. Long lines of white breakers move swiftly up the sand or throw up fountains of foam against the rocks. Then the waterdrops fall back into the rapidly receding surf before the next wave comes along. But not all the water returns immediately to the sea. Some tiny droplets are carried upwards into the air and begin to evaporate. In so doing they leave behind minute salt particles, which play a vital role in the formation of clouds and rain. As we shall see, water vapor condenses on particles such as these. They serve as nuclei on which the cloud droplets grow.

There are also other types of nuclei. Some are better than others because they have a higher affinity for water molecules and grow into cloud droplets at humidities below 100 per cent. Fires of all kinds are producers of nuclei; in consequence, large cities make a contribution to the nuclei population of the atmosphere. Nowadays people all over the world are worried about air pollution. The term "smog," which is a combination of the words smoke and fog, has been used often enough to be included in some dictionaries. Although there is still some disagreement as to the major sources of the smoke, there is no doubt that increasing quantities of gases and particles are being dumped into the atmosphere from smokestacks, automobile exhausts, and back-yard fires. As you approach a large city, you often see a dark brown pall over the skyline. It is sometimes called the "dust dome." Without question, air pollution has reached the stage of a major problem. Not only does it mar the beauty of a city, but the pollutants in the air may cause extensive damage to buildings and vegetation, and, more important, they represent a serious hazard to human health.

One can hardly be happy about the smoke pouring out of the stacks of a large factory, but it should be recognized that the particles may serve as nuclei for cloud droplets. Would there be less rain on earth if all the furnaces were suddenly extinguished? The answer is that probably no effect would be noticed. Long before the growth of heavy industry, the automobile, and the rapid increase of air pollution, the quantity of rain and snow was not much different from what it is today. Nature accommodates herself to the particles which are available. In the "old days," lightning, forest fires, and volcanoes occurred, just as they do today. Also, certain types of soil particles may serve as nuclei, and there has been little change in this source over the decades.

What is the role played by the nuclei? Exactly how are they produced? How large are they? Let us look into these questions in the next few pages.

Why Is a Nucleus Necessary?

Everyone knows that on a warm, humid day the outside of a glass of ice water becomes very wet in a matter of minutes. It is also well known that this wetting occurs because water vapor from the air condenses on the cold glass surface. The air surrounding the glass is cooled. As this cooling goes on, the relative humidity increases, the relative humidity being the ratio of the amount of water vapor present in the air at a given temperature to the amount of water vapor the air could hold at the same temperature. When the relative humidity reaches 100 per cent, the air is said to be saturated. Condensation begins. What would you expect to occur if the air were cooled in the absence of solid objects like glass surfaces? Many experiments along these lines have been conducted, and we now have a good idea.

If one takes great caution to clean the air of all types of particles, including the minute, electrically charged ones called ions, then it is possible to increase the relative humidity to very high values before condensation of water vapor occurs. As a matter of fact, a concentration of water vapor several times the saturation amount is required before condensation occurs in the form of tiny cloud droplets. When the air contains more water vapor than the quantity it would have at saturation, it is said to be *supersaturated.* In air absolutely free of particles we find supersaturations of "several hundred per cent," meaning that the air contains several times more water-vapor molecules than it would at saturation. The reasons why such great supersaturations are needed before droplets form have been explored theoretically by a number of scientists.

When we say that condensation has formed a water droplet, we mean that a large number of water molecules have come together to form a liquid. When there is no foreign surface, such as the surface of a cold glass, or a nucleus, the water-vapor molecules come together only as a result of accidental collisions. Groups of molecules are constantly being formed and then broken up as the molecules fly off in all directions. It has been shown, however, that if a sufficiently large number of molecules can be brought together, they form a particle which continues growing instead of rapidly evaporating. Instead of speaking of a "sufficiently large number of water-vapor molecules" one can consider the molecules to be aggregated into a sphere. It then becomes possible to speak of the radius of the sphere of molecules which must be exceeded in order to produce water droplets. In 1870, Lord Kelvin, the famous British scientist, first developed an equation which permits calculations of the critical radii. They depend on the temperature of the air as well as the degree of supersaturation. The higher the supersaturation, the greater the likelihood that the critical number of water-vapor molecules will aggregate to form a droplet which will continue growing.

Theoretical work, supported by laboratory experiments, clearly shows that when there is no surface on which the air may condense, very high supersaturations are needed before droplets will form. In the atmosphere supersaturations of only 1 or 2 per cent (rather than several hundred) may sometimes be reached in rapidly cooling air, but still clouds are common. The reason is, obviously, that in the atmosphere there are always nuclei on which water-vapor molecules may collect in sufficient numbers to form droplets.

We have already mentioned sea-salt nuclei. The fact that water vapor condenses readily on salt is common knowledge. You undoubtedly have tried to shake salt on your food on a humid summer day. The salt particles in the shaker absorb water vapor and agglomerate and cannot pass through the holes, shake as you will. Once a salt nucleus becomes wet, it dissolves in the water and forms a solution, which attracts more water vapor and continues growing.

The salt particles in the atmosphere range in size from less than 0.01 micron to as large as 10 microns.*

* A micron is equal to one millionth of a meter and 0.0000394 inch. See the Appendix for a list of units commonly used by meteorologists and for conversion factors from English units to metric units.

Techniques for capturing these small particles and measuring them are discussed in the next section.

There is another large group of nuclei which have been studied in detail by Christian E. Junge, formerly of the Air Force Cambridge Research Center. These droplets contain chemical substances known as sulphates. They are produced in the atmosphere as a result of burning substances containing sulphur. When coal, for example, is burned in a furnace, the smoke will contain sulphur-dioxide gas, formed by a combination of sulphur and oxygen. When this gas is exposed to oxygen, it is converted to sulphur trioxide. When this trioxide, in turn, is exposed to water vapor, it becomes sulphuric acid, H_2SO_4. The exposure of the gases to sunlight accelerates the conversion of sulphur dioxide to sulphuric acid. Junge has shown that many nuclei are composed of ammonium sulphate.

Tiny nitrous acid droplets are formed by the combination of nitrogen, oxygen, and water vapor in the presence of high temperatures. These conditions accompany forest fires and, particularly, lightning storms. Some industrial furnaces also contribute nitrous acid to the atmosphere.

Minute acid droplets act as condensation nuclei because, like salt, they are hygroscopic—that is, they have an affinity for water. Water vapor begins condensing on them at relative humidities below 100 per cent. Nuclei of this type are less than a few tenths of a micron in size and are found in large quantities all over the world.

Soil particles carried from land surfaces by winds make a third major source of condensation nuclei in the atmosphere. Those particles with diameters greater than about 10 to 20 microns fall back to the earth rapidly because they are heavy, but the smaller particles may be transported to high altitudes and over great distances. The efficiency of the soil particles as condensation nuclei depends on their properties. The most important ones again are those that are hygroscopic, that is, those which are easily wet by water and easily dissolved in water.

It is convenient to classify nuclei into three groups (Fig. 1). The smallest ones have diameters less than 0.4 micron and occur in concentrations of 1000 to 5000 per cubic centimeter (cm^3). They are commonly called *Aitken nuclei* after the scientist who, in 1880, demonstrated that water vapor condenses on nuclei. The second group are called *large nuclei.* They have diameters between 0.4 and 1 micron and have concentrations of a few hundred per cm^3. *Giant nuclei,* the largest observed, have diameters between 1 and 10 microns and occur in concentrations of 0.1 to $1/cm^3$. For the most part, large nuclei are composed of sulphates, while giant nuclei are mostly sea-salt particles.

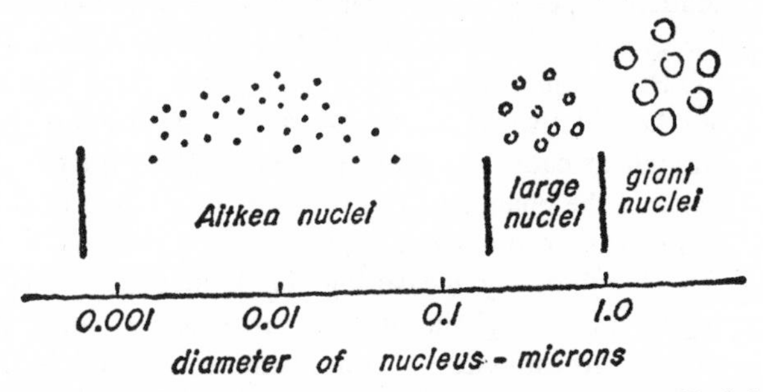

FIG. 1. Condensation nuclei are conveniently classified in three size groups.

MEASURING THE SIZES AND NUMBERS OF NUCLEI

Over the years some ingenious schemes have been devised for studying the properties of condensation nuclei. When you realize that most nuclei are too small to be seen by the naked eye and that some cannot be seen with an ordinary microscope, it is obvious why ingenuity has been necessary.

The properties of the nuclei in which we shall be interested are the sizes, the number in a unit volume of air, and their compositions. We have noted already that salt particles, acid droplets, and soil particles are commonly found in the atmosphere. There are many other types of particles, but these three groups are of greatest interest to the cloud physicist.

It would be instructive to examine briefly some of the techniques employed in the past. One of the simplest devices for measuring the number of condensation nuclei per unit volume, called the *concentration,* is the so-called Aitken nuclei counter. It consists of a vessel which is sealed shut as the pressure inside is increased above atmospheric pressure. The chamber is then permitted to come to room temperature, and a valve is suddenly opened. As the air rushes out of the chamber, the pressure drops suddenly. The expansion leads to a sudden temperature drop and causes a rapid increase of the relative humidity to a supersaturation large enough to cause condensation on virtually all the particles in the chamber. The presence of a cloud of droplets can be seen when a beam of light is passed through the chamber. By measuring the reduction of the intensity of the light as it goes through the cloud, it is possible to estimate the concentration of droplets. This information gives the concentration of nuclei, since each droplet forms on a nucleus.

The degree of supersaturation in an Aitken nuclei counter depends on the amount of expansion. If very minute nuclei are to be detected, a high degree of supersaturation must be produced. Unfortunately, the Aitken counter gives no information about the size or composition of the particles. It has shown, however, that concentrations of minute condensation nuclei are about five times higher over land than over oceans. This result indicates that the major sources must be over continents rather than oceans.

Now, in order to examine individual nuclei, it is necessary to capture them. How in the world, you may ask, does one go about capturing a particle 1 micron in diameter?

If you had a filter with sufficiently small holes, you could pass air through it and strain out the particles. James P. Lodge, Jr., has used filters with pore sizes of about 0.3 micron to capture very tiny particles. Another technique involves the impaction of the particles on a surface. A common impactor design is shown in Figure 2. Air is forced through a nozzle so that its speed at the end of the nozzle is very high. A glass or plastic slide is placed a very short distance, perhaps a millimeter, from the nozzle. As the air containing the particles impinges on the slide, it is deflected outwards. On the other hand, if the particles are large enough, they cannot be deflected enough to escape contact with the slide, and they strike the slide and stick to it. To insure sticking, the slides are sometimes coated with a tacky substance. The impaction devices can effectively capture particles having diameters larger than about 1 micron. Smaller particles closely follow the air stream and do not impact on the slide.

The techniques of identifying the particles, once they have been captured, are quite clever. The tacky substance already mentioned may be a special gelatin impregnated with a chemical that reacts with a particular type of particle. Benjamin K. Seeley, at the New Mexico Institute of Mining and Technology, was interested in particles containing chlorides because he was interested in the chemical compound sodium chloride, common salt. He mixed a chemical called mercurous fluosilicate in the gelatin. When a salt particle was impacted into the gelatin, it immediately began to react with the reagent and produced a spot which had a distinctive blue appearance when viewed under a microscope. Tests showed that the blue spot grew to a diameter about nine times larger than that of the original salt particle. This property made it easy to identify the places where salt particles had landed, to count the total number on the slide and to measure the size of each particle.

The filter technique developed by Lodge also uses

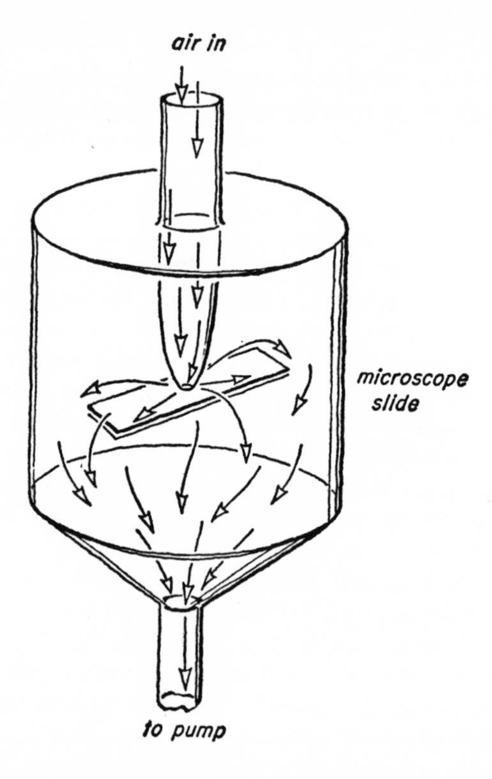

FIG. 2. An impactor for collecting particles of diameters greater than about one micron. As the air moves through the nozzle at high speed, the particles impinge on the microscope slide and stick to it. The slide is normally placed about 1 millimeter from the end of the nozzle.

the reagent mercurous fluosilicate. The filters are floated in a solution of this reagent so that the particles of salt may react and leave the characteristic blue spots.

Ottavio Vittori, in Italy, also has employed the gelatin scheme. However, he impregnated gelatin with silver nitrate. When chloride particles fall on the gelatin, they produce an almost perfect ring. When the slide is exposed to light, the ring, viewed under a microscope, is quite distinct. At first it has a yellow color, but it gradually turns brown.

A fascinating technique was first employed by Henri Dessens, in France. He used spider webs. To capture very tiny particles by impacting them on a surface, it is necessary either to accelerate them to high speeds or to use a very narrow collecting surface. Dessens made use of a species of spider that produced very uniform threads, about a hundredth of a micron in diameter. After a spider was captured, it was permitted to spin thread around a frame. The network of threads was exposed to a known quantity of air and then placed in a special chamber under a microscope. Although examination of the particles caught on the threads sometimes showed they were dry solid particles, most of them were found to be tiny droplets even when the relative humidity was as low as 50 per cent. As the humidity was reduced, the water evaporated until minute white cubic crystals were left. They were identified as crystals of sodium chloride. Dessens detected particles with diameters as small as 0.5 micron.

Investigators also have sent their particle-collecting devices aloft in airplanes. The most extensive flight measurements of sea-salt nuclei have been made by Alfred H. Woodcock and his colleagues at the Woods Hole Oceanographic Institution. They used glass and silver rods about 0.5 mm in diameter and had them exposed to the air streaming past the airplane. Particles that struck the rods were collected. The rods were placed in a chamber whose relative humidity and temperature could be controlled. By viewing the rods through a microscope the investigators could see that most of the "particles" were actually tiny water droplets. Knowing the temperature and humidity of the air in the chamber, they could infer, from a knowledge of the sizes of the droplets, the sizes of the salt particles. Because the diameter of the rod was relatively large, this scheme could not detect nuclei smaller than about 1 micron in diameter.

Many other devices are variations of the filter or impaction techniques. In some the flow of air is such as to cause a sorting of the particles according to size (see Fig. 3). In others the sorting is accomplished by shaping the impacting surface in the shape of a curve of increasing curvature. The results of such a

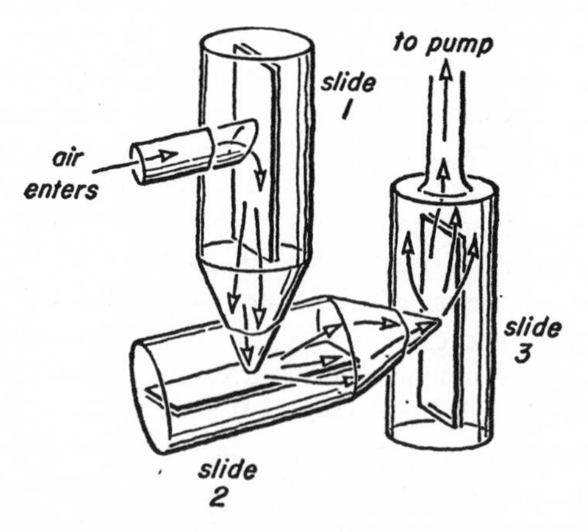

FIG. 3. One type of cascade impactor. Air moving through the collector passes through successively smaller nozzles at successively greater speeds. The succeeding slides collect smaller and smaller particles. Slide 1 collects the largest particles, slide 2 smaller particles, and slide 3, the smallest ones.

design are to cause the large particles to be impacted first, and the smaller particles to be impacted downstream.

For capturing extremely small particles–those from 0.01 to 0.1 micron–so-called thermal precipitators have been employed (Fig. 4). Two plates are fixed parallel to one another and quite close together. One plate is heated while the other is cooled. Air close to the hot plate has a higher temperature than the air close to the cool one. The air molecules in the warmer air move about more violently and, through collisions with the tiny particles, drive them towards the cooler plate. By placing a microscope slide on the cool plate one can capture the particles. In some studies slides for electron microscopes have been used so that the particles can be enlarged many thousands of times before viewing. Miss Barbara Tufts, at The University of Chicago, developed tests for identifying many types of particles by means of the electron microscope. Notwithstanding the extremely small sizes, atmospheric particles can be captured, identified, and counted, and

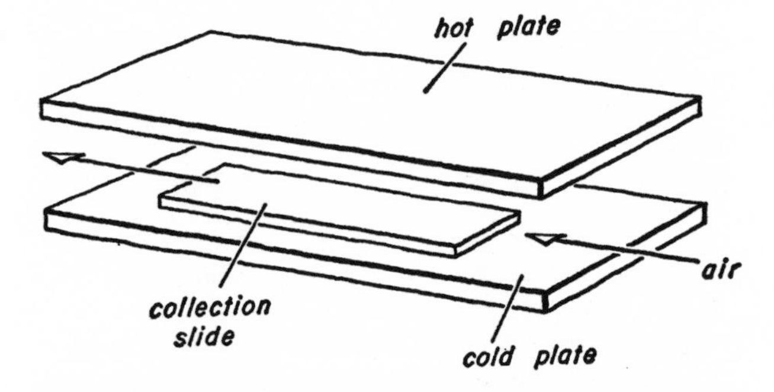

FIG. 4. Schematic drawing of a thermal precipitator for the collection of very small particles, those with diameters smaller than one micron. The tiny particles are driven toward the cold plate and land on the collection slide.

it will be instructive now to get an idea of the sizes and the numbers actually observed.

Some Condensation Nuclei Statistics

As already noted, scientists started making measurements of condensation nuclei more than sixty years ago. The early observations of Aitken and others gave only the number of nuclei per unit volume in the formation of water droplets at very high supersaturation. Little was learned about the properties of the particles. In the last twenty years refined techniques have made it possible to obtain information about the numbers, sizes, and composition of the particles.

All types of observations have revealed that there is a great variation in nuclei contents from one time and place to other times and places. In 1938, H. E. Landsberg, now with the U. S. Weather Bureau, compared measurements taken of Aitken nuclei in a variety of places. The results are shown in Table 2. This table clearly illustrates that in cities the numbers of nuclei

TABLE 2. COMPARISON OF AITKEN NUCLEI IN VARIOUS PLACES
(*From H. E. Landsberg*)

	No. of observations	Average concentrations (per cm^3)	Range of concentrations (per cm^3)
City	2,500	147,000	3,500 to 4,000,000
Country ...	3,500	9,500	180 to 336,000
Ocean	600	940	2 to 39,800

are much higher than in the country and very much higher than over oceans. The actual concentration at any time in any locality depends on many factors: for example, the types of industry, types of soil, wind velocities, whether or not it has rained recently. However, the statistics presented here show that the major sources of Aitken nuclei must be over continents and that human activity increases the number in the air.

It should again be noted that Aitken nuclei are generally so small that high supersaturations are needed to cause condensation to occur on them. This is clearly brought out by noting that the normal cloud-droplet concentrations range from about 10 to 1000 per cm^3, while the Aitken nuclei sometimes are over 100 times more numerous. Only the largest and most efficient Aitken nuclei are ever involved in cloud formation.

The concentrations of large nuclei are more nearly equal to the number of cloud droplets per unit volume. Table 3 presents some observations obtained by Dessens with his spider webs. The network of threads was exposed to a known volume of air, and the size and number of particles were noted when the spider web was exposed to an atmosphere with a relative humidity of 78 per cent. Even in the open country he found variations from day to day, but the ranges of values were very much smaller than for Aitken nuclei.

TABLE 3. NUMBER PER CUBIC CENTIMETER OF SMALL DROPLETS CONTAINING CONDENSATION NUCLEI
(From H. Dessens)

Radius in microns	*Number*
Below 0.1	100
0.2	46
0.3	30
0.4	14
0.5	7
0.6	2
0.7	1
Total	200

It is clear from Table 3 that the smaller the particle, the greater the number in a known volume of air. Particles greater than about 1 micron are considered giant nuclei. Although they occur in concentrations smaller than $1/cm^3$, they are very important in the rain-formation process.

We mentioned in an earlier section that extensive measurements of sea-salt particles have been made by Alfred H. Woodcock. Figure 5 shows a diagram taken from one of his reports. It presents the concentration of giant nuclei compared with the radius of the nucleus. The radius here corresponds to that which the particle

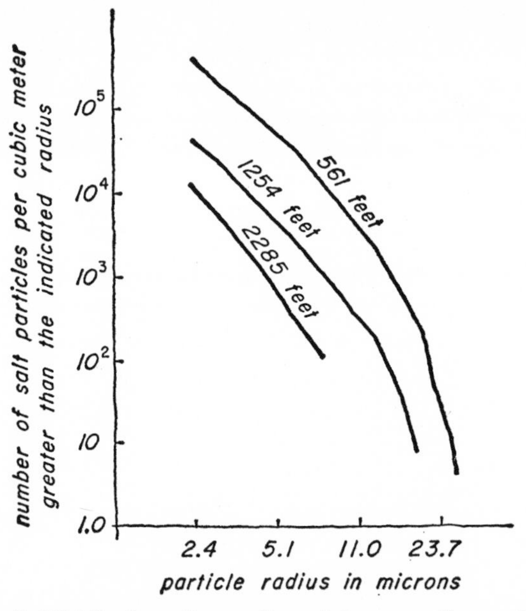

FIG. 5. Distribution of sea-salt particles according to their radii and altitudes. Curves show how the number of particles per cubic meter whose radii are greater than the size given on the horizontal scale varies at three altitudes. The sizes were measured when the particles were suspended in a chamber where the relative humidity was 99 per cent. Measurements were made over the ocean near Hawaii at three altitudes. The numbers on each curve give the altitudes in feet. The lowest altitude was below the cloud bases, the middle altitude in the cloud layer, and the highest altitude above the tops of cumulus clouds. (From Alfred H. Woodcock)

would have at a relative humidity of 99 per cent. In such a humid atmosphere the salt particles will be in the form of a solution of salt water.

As is true of virtually all particles in the atmosphere, the number per unit volume decreases as the particle size increases. The number of 5-micron particles ranges from 10^3 to $10^5/m^3$ (i.e., from 0.001 to $0.1/cm^3$). Thus, in a typical cloud there will be relatively few giant nuclei compared to the number of cloud droplets.

It is well established that the oceans are the chief sources of giant salt nuclei. Woodcock has made observations over the ocean under many different conditions. He found that, in general, the higher the wind velocity the greater the concentrations of nuclei. This result is easily explained. The higher the wind speeds the greater the wave action, and the greater the number of droplets of sea water thrown into the atmosphere. Woodcock and others also have made observations of sea-salt nuclei at various altitudes. As you would expect, the greater the altitude above the ocean source the smaller the concentration of sea-salt nuclei. Very recently Woodcock and an associate, A. T. Spencer, have shown that when hot lava from Hawaiian volcanoes falls into the ocean, great quantities of sea-salt particles are ejected into the atmosphere.

Over continents the concentrations of giant nuclei decrease because they fall out of the air or are washed out by falling rain. Nevertheless, giant salt nuclei are sometimes found far inland. Horace R. Byers and his colleagues at The University of Chicago made flight observations over the central United States. On some days they found salt nuclei greater than 5 microns in diameter in concentrations of 100 to $1000/m^3$.

In summary, it should be noted that the atmosphere is full of particles of many types ranging in diameter from about 10 microns to below 0.01 micron. In general, the smaller the sizes the more numerous the particles. Typical clouds have concentrations of water droplets ranging from 10 to 1000 per cm^3. Only a small fraction of the available nuclei actually take part in cloud-droplet formation. The chief factors determining whether a particular particle will act as a cloud-droplet nucleus are the particle size and its composition. A large particle composed of a substance having an affinity for water is in a favored position. The manner in which cloud droplets are formed is discussed in the next chapter.

Chapter 3

CLOUDS AND THEIR FORMATION

For centuries clouds have had special meanings to all kinds of people. Poets and painters have found inspiration in the beauty of their forms and colors. Farmers have looked up to them gratefully when they produced rain to feed thirsty crops. Other farmers have shaken their fists at the clouds when they sent down hail or strong winds which destroyed grains or fruits ripe on the vine. Meteorologists look on clouds with less emotion, but what they see is no less interesting.

What Is a Cloud?

A cloud to a meteorologist is a collection of minute water or ice particles which are sufficiently numerous to be seen. The size of cloud droplets ranges from a few microns to as large as 100 microns. Plate I shows a photograph of a group of cloud droplets captured on a microscope slide. In this cloud most of the droplets were about 10 to 20 microns in diameter. As will be seen in a later section, many clouds have large numbers of droplets of this size.

The cloud droplets shown in the photograph are almost perfect spheres. In a later chapter it will be shown that as waterdrops grow to larger sizes and become raindrops, their shapes depart more and more from the spherical. You might ask what the difference is between a cloud drop and a raindrop. The major difference is one of size. The larger the size, the faster the speed of fall of the drop and the greater the distance it will fall before it evaporates. These points are clearly shown in Table 4. The information on evaporation rates was

TABLE 4. FALL SPEED AND DISTANCE OF FALL BEFORE EVAPORATION OF VARIOUS WATERDROPS

Drop radius in microns	Fall speed in cm/sec	Distance of fall before complete evaporation in meters
10	1	Less than 1
100	76	150
1,000	690	4,200

calculated by a German scientist, W. Findeisen, on the assumption that the drops were falling in air whose relative humidity was 90 per cent. Drops of radii smaller than 100 microns fall very slowly, and when they descend out of a cloud they evaporate very quickly. On the other hand, drops of radii greater than 1000 microns fall quite fast and may descend several kilometers below the cloud base before evaporating. As a consequence they may reach the ground in the form of rain. It has been decided that it would be convenient to use a radius of 100 microns as the demarcation between cloud droplets and rain droplets.

CAPTURING CLOUD DROPLETS

Over the years many studies have been made of the sizes and numbers of cloud droplets. A variety of schemes has been developed for capturing them. The most successful ones are quite simple. A microscope slide is covered with a thin layer of oil or vaseline. The slide is mounted at the end of a rod which is taken aloft by airplane and exposed to the cloud air as the plane flies along. When the cloud droplets strike the slide, they sink into the oil and are examined under a microscope. For permanent records a camera is fixed on the microscope. The photograph shown in Plate I was obtained in this way.

By noting the time that the slide is exposed and the speed of the airplane, you can calculate the volume of the cloud air sampled. The method then permits you to measure the number of droplets per cubic meter or cubic yard or whatever other volume unit you may choose.

One of the inconveniences of using oil-coated slides is that they must be examined under the microscope immediately after collection. Even though the water droplets are submerged in oil, they evaporate fairly rapidly because they are so small. An observer must keep his eye fastened to the end of a microscope tube in order to focus the microscope before the camera is snapped. If the air is turbulent, as it frequently is in summer clouds, this occupation is hard on observers. If the airplane suddenly lunges upward, the observer is likely to go home with a black eye.

Instead of employing oil, some scientists have used smoke to coat the slides. A smoke of very fine magnesium-oxide particles is produced by burning a ribbon of magnesium. When a glass slide is slowly passed through the smoke, a thin, uniform film of particles is deposited on the slide. The size of the individual particles is about 0.5 micron. For the purpose of cloud-droplet sampling, properly coated slides should have a thickness of smoke particles greater than the largest cloud droplets expected to be in the cloud.

When the smoke-coated slide is exposed to a cloud as an airplane flies through it, the droplets sink into the soft layer of smoke particles. When they evaporate, they leave round, permanent holes. The slides can then be stored and studied after the airplane has landed and a suitable microscope is available.

Some researchers have developed cameras to photograph cloud droplets directly through a microscope. In order to obtain the proper focus it is necessary that a very small region be viewed through the microscope tube. The result is that each photograph contains no more than a few droplets. Since a typical cloud contains perhaps 200 droplets per cubic centimeter, one must examine thousands of droplets (that is, thousands of photographs) in order to obtain a representative picture of the droplet sizes in the cloud. This fact has made direct photography of cloud droplets impractical.

There are many other techniques for sampling cloud droplets. Some are more successful than others, but none is more reliable than those involving the capture of droplets on a slide. In recent years a number of schemes have been devised which make use of the fact that the scattering of light waves by a drop depends on the size of the drop. If clouds were composed of droplets all of the same size, practical devices which make use of light-scattering principles could be devised. Unfortunately, from this point of view, the droplets in any cloud are not uniform but vary over a large range.

One new device under development offers some hope of solving this problem. It involves the examination of the light scattered by individual cloud droplets as they pass through a very narrow beam of light. The scattered light is captured and recorded electronically. True, only one droplet is examined at a time, but the number of droplets which can be observed in a single minute can be in the thousands. Unlike the other schemes, which require that each droplet be examined and its size measured manually, this new device can automatically record the number of droplets falling into various size intervals. A series of Russian articles published between 1956 and 1959 says that a workable device of this type has been developed. However, until it has been tested in practice, we can only be hopeful. The history of observational instruments of this type shows that they often work well in a laboratory but fail when taken into the field.

Sizes and Numbers of Cloud Droplets

Over the years everyone sees many types of clouds, ranging from huge thunderclouds to shallow layers of fog. Although fog at the ground is usually not thought of as a "cloud" in the common sense of the word, it has most of the properties of a cloud. Sometimes you observe a very low layer of dark, uniform cloud. It is called *stratus* and can be considered a fog which has been lifted a few hundred feet above the ground.

The spectrum of cloud-droplet sizes varies from one cloud type to the next, from one cloud of one type to another cloud of the same type. As a matter of fact, from one place and time to another place and time in the same cloud, you often find large differences in the droplet characteristics. But even with all these variations, one can show that, on the average, the droplets in certain types of clouds are considerably different from those in other clouds. Figure 6 shows how the number of droplets of various sizes differs in three types of clouds in the cumulus family.

Cumuli of fair weather are small, white puffy clouds. They seldom are more than 3000 feet, and often are less than 1000 feet, thick. They never produce rain. It can be seen that in these types of clouds the droplets are small and numerous. The maximum diameters in most of them are less than 50 microns and the concentration of droplets is above $300/cm^3$.

In large cumulus clouds, which produce showers over the central United States, it is found that the droplet diameters often exceed 50 microns. The concentrations of droplets are smaller than those in fair-weather cumuli—they run around $200/cm^3$.

The cumulus clouds over tropical oceans have droplet characteristics differing markedly from those found in continental clouds. The clouds themselves differ in many respects. In particular, they produce rain when they are quite small. Observations from airplanes show that the cloud bases usually are at an altitude of about 2000 feet, and when the cloud tops exceed 8000 to

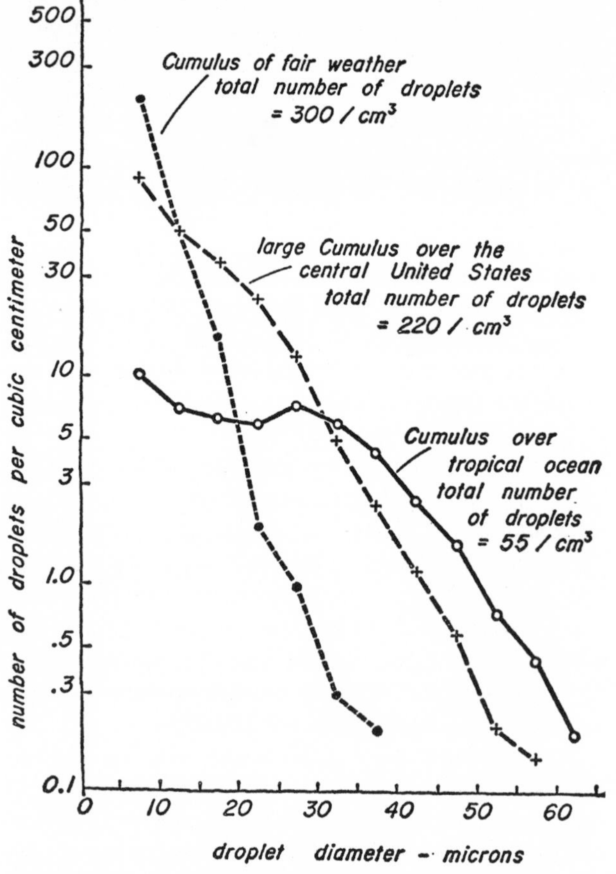

FIG. 6. The spectrum of droplet sizes in three types of clouds. Each point on each curve represents the number of droplets whose diameters are within a diameter interval of five microns. The total number of droplets per cubic centimeter is shown with each curve.

10,000 feet, rain is often seen to fall. Cumulus clouds over continents seldom produce rain unless their summits rise as high as 15,000 to 20,000 feet. The cloud droplets in tropical clouds occur in concentrations of only about 60/cm^3, but there are usually some droplets of diameters exceeding 50 microns. It has been suggested that the presence of many large cloud droplets in shallow clouds can be attributed to a large number of giant salt nuclei. We shall say more about this later.

In layer-type clouds the droplets are usually smaller than in cumuliform clouds. Measurements by a number of investigators have shown that in most stratified clouds the average droplet radii range in sizes from about 4 to 10 microns, while the concentrations range from about 200 to 600/cm^3. In shower clouds the average radii are near 20 microns, and concentrations are 50 to 200/cm^3.

The Growth of a Cloud Droplet

Everyone is familiar with the process of condensation. We have already mentioned the familiar experiences with a glass of ice water. On cold days you can clearly see the process at work by merely breathing out and watching clouds of tiny water droplets form. The water vapor is supplied in the air exhaled from your lungs. There are many other examples: the white cloud above the steam locomotive, the pearly white trails produced by high-flying jet airplanes.

Condensation is a process whereby water-vapor molecules are caused to come together in sufficient numbers to produce liquid water. When large surfaces are concerned, we need only consider the nature and temperature of the surface and the temperature and relative humidity of the air in order to understand how condensation comes about.

Consider the glass of ice water on a day when the

relative humidity of the air is 50 per cent and the temperature is 80°F. When the glass of water is first set on a table, the air in contact with the glass begins to cool because the temperature of the glass is that of a water-and-ice mixture, that is, 32°F. As the air cools, its relative humidity increases even if no water vapor is added to it.

In order to understand why this happens, let us digress for a moment and examine the properties of air and water vapor. Consider a closed jar with a small amount of water in the bottom. If the air inside the jar had a relative humidity of 50 per cent when the water was added, the water would begin to evaporate; water-vapor molecules would escape from the liquid to the air. In reality, some of the water-vapor molecules in the air also enter the water, but they are fewer in number than the number of molecules going from the water to the air. As time progresses, the number of water-

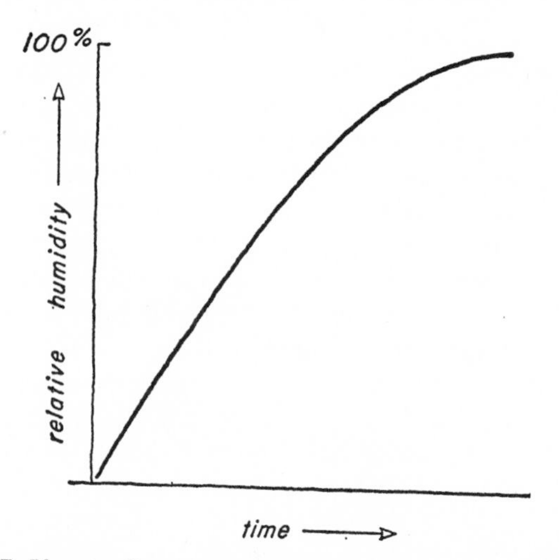

FIG. 7. If a small amount of water were added to a jar of dry air, the relative humidity of the air would increase with time in the manner illustrated by the curve.

vapor molecules in the air increases, and the relative humidity rises. As it does, the difference between the rate of transfer of molecules from water to air and from air to water decreases. This in turn leads to a decrease of the rate at which the relative humidity of the air is increasing. Figure 7 shows how the relative humidity increases with time. When the curve reaches the 100 per cent level, the air is said to be *saturated.* At this point there is an *equilibrium condition;* the number of water-vapor molecules going from water to air exactly equals the number going from air to water. One can measure the quantity of water vapor in the air by measuring the pressure of the water vapor. At saturation, the pressure is called the *saturation vapor pressure.*

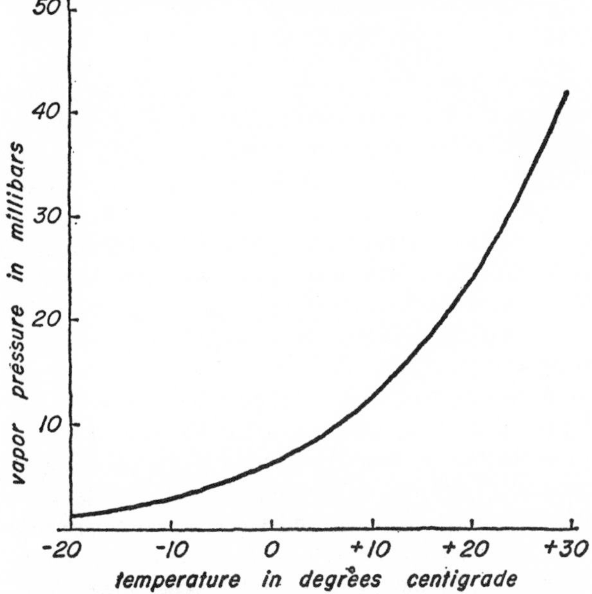

FIG. 8. The vapor pressure of air saturated with water vapor depends only on the temperature of the air. The curve shows how the saturated vapor pressure increases as the temperature increases.

It is well known that the quantity of water vapor which can be supported by air depends on the temperature. As the temperature rises, the saturation vapor pressure increases (Fig. 8).

In the experiment just discussed it was assumed that the temperature was maintained at a constant value. If, after the air has reached the final saturation value, the jar were put in a refrigerator so as to lower the temperature, it would be found that water-vapor molecules in the air would begin to migrate back into the water. At the lower temperatures the saturation vapor pressure of the air is reduced, and as a result the number of water-vapor molecules which the air can support is diminished.

The Dew Point

Let us now return to the example of the glass of ice water. As the air surrounding the glass is cooled, the relative humidity is increased. In a short time the air reaches a relative humidity of 100 per cent and is saturated. Further cooling means that the air will become supersaturated; that is, it will have more water-vapor molecules than in the stable saturated condition. Water-vapor molecules will begin to deposit themselves on the surface of the glass in order to restore the air to the saturated state. As long as the temperature continues to fall, there will be a continuation of the condensation process. Of course, in the case of the ice water the humid air in the vicinity of the glass is being constantly replaced. Consequently, the condensation process goes on and on, and the glass continues dripping water on the tablecloth. If there were a limited supply of air, the process would slow down as the temperature of the air surrounding the glass decreased.

The point in the cooling process at which condensation just begins is called the *dew point*. It occurs when the relative humidity reaches 100 per cent. The temperature of the air at which condensation just begins is the *dew-point temperature.* This very convenient figure depends on the air temperature, moisture content, and pressure. In the case already cited—namely, when the temperature is 80°F and the relative humidity is 50 per cent—the dew-point temperature at standard atmospheric pressure is 60°F.

On a fairly clean metal or glass surface condensation begins when the temperature of the surface equals the dew-point temperature of the air. In the case of very minute surfaces composed of substances having an affinity for water vapor the situation is more complicated.

First of all, when very tiny droplets (with radii less than about 1 micron) are involved, the shape of the droplet affects the saturation vapor pressure over the drop. In order for such small droplets to be in equilibrium with the surrounding air, it is necessary for the relative humidity to be greater than 100 per cent. For example, if only pure water were involved, it would be necessary to have relative humidities of over 140 per cent in order to prevent a droplet of radius 0.003 micron from evaporating immediately. As the droplets increase in size, the equilibrium relative humidity approaches 100 per cent.

In the atmosphere relative humidities seldom exceed 101 per cent, and even these 1 per cent supersaturations occur only in very strong updrafts in thunderstorms. How, then, can cloud droplets form? The answer is that the tendency of very tiny droplets to evaporate is counteracted by the affinity of certain substances for water molecules. A particle of sea salt has a great fondness for water. As already noted, over a clean metal or glass surface a relative humidity of 100 per cent is needed before condensation begins. In the case of a sea-salt or other hygroscopic particle, condensation can occur with relative humidities of only 50 or 60 per cent.

The role played by a salt nucleus in the growth of a cloud droplet is shown in Figure 9. The dashed curve represents the relative humidities, measured with respect to a plane water surface, that are needed to maintain in equilibrium pure water droplets having the radii shown on the horizontal scale. It can be seen that as the droplets' sizes decrease, higher supersaturations

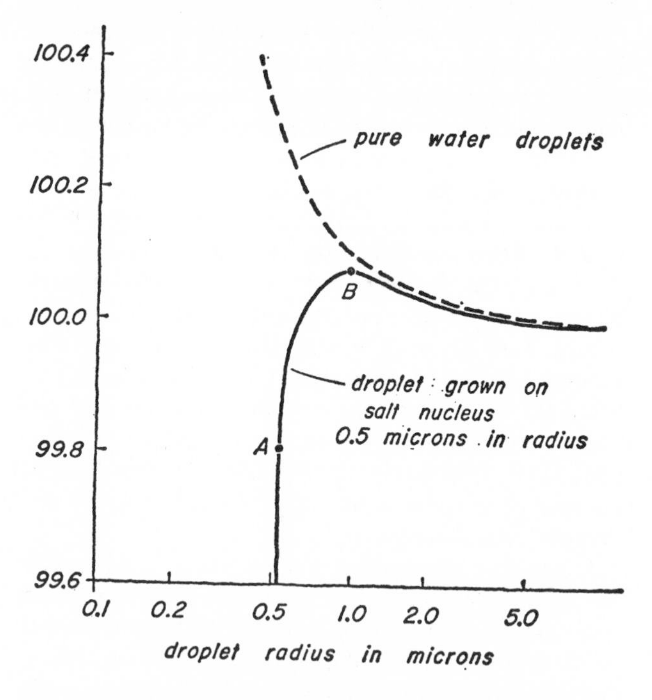

FIG. 9. The equilibrium relative humidity—that is, the relative humidity at which a droplet, whose radius is given by the horizontal scale, remains the same size—depends on the size and the composition of the droplet. If the actual relative humidity is greater than the value shown on the curve, the droplet grows by condensation. If the relative humidity is less than that given by the curve the droplet evaporates. These curves are valid with a temperature of 0°C. See text.

are needed in order to prevent the droplets from evaporating.

The solid curve shows the relative humidities needed to maintain equilibrium over a droplet which started growing on a salt nucleus of 0.5-micron radius.

When a salt nucleus of this size is introduced into an atmosphere of relative humidity 99.8 per cent, condensation will occur on the nucleus until it grows to the size indicated by point A. At this time the nucleus will be in equilibrium with its surroundings and remain this size unless the relative humidity is changed. If it were increased to 100.1 per cent the condensation and droplet growth would resume. Once the droplet radius "passes over the hump" (point B) in the curve, it will continue to grow as long as the relative humidity exceeds 100 per cent. By the time the droplet has grown to a size greater than about 2 microns, the size effects and salt-solution effects are so small that the droplet acts like a large surface of pure water. When this happens, we have a situation similar in some respects to that of the glass of ice water. Condensation continues as long as the air is supersaturated with respect to the pure water.

Up to this point we have specified the behavior of the droplets in terms of the relative humidity of the surrounding air. You can more clearly understand the process of condensation if you think about this formula for relative humidity:

$$\text{R. H.} = \frac{\text{Actual vapor pressure}}{\text{Saturation vapor pressure}} \times 100$$

We have mentioned already that the vapor pressure is a measure of the quantity of water-vapor molecules in the air. One can also speak of the vapor pressure at the surface of a liquid; it is equal to the saturation vapor pressure of the air just at the surface of the liquid and depends only on temperature. When the air over a water droplet is supersaturated—that is, when its relative humidity is greater than 100 per cent—the vapor pressure of the air is higher than the saturation vapor pressure. As a result, there is a pressure force driving the water-vapor molecules from higher to lower pressure—namely, from the air to the water. This is the condensation process at work.

Condensation and evaporation of droplets of any size, shape, and composition can be calculated if it is possible to describe the field of vapor pressure around the drop.

In summary, we can say that the growth of an individual cloud droplet depends on the size and composition of the nucleus and the moisture of the air. When the relative humidity of the air is increased to sufficiently high values, condensation of water vapor begins. If the relative humidity is maintained at a value slightly greater than 100 per cent, condensation continues and a cloud droplet may be formed.

At this point the inquiring reader may ask various pertinent questions. What causes large bodies of air in the atmosphere to cool sufficiently to produce extensive cloud masses? What happens when millions of nuclei are activated almost simultaneously in a cooling body of air? These and other questions are considered in the next section.

THE GROWTH OF A CLOUD

Nature has provided various means to cool air below its dew point so that clouds may form. On clear nights great quantities of heat are radiated outward from the lower atmosphere. When the layers of air near the surface are moist while the upper layers are dry, there is pronounced cooling of the earth's surface and of the low-lying moist air. In certain circumstances the cooling continues until the dew-point temperature is reached. When this occurs, fog forms. You probably have noticed that fog near the ground often forms first in low spots–in dips in the road, in valleys, and so forth. The reason is that the cooler air is heavier and sinks into the depressions in the topography.

Fogs are also produced when warm moist air passes over cold bodies of water. Here the air is cooled by a loss of heat to the water. If the air temperature is lowered to the dew point, condensation and fogs occur.

Still another type of fog is produced by the passage of cold air over a warm body of water. Condensation occurs because water vapor is added to the cold air in sufficient quantities to saturate it. This is similar to the effect produced by a steam engine.

Although the processes of increasing the relative humidity which lead to fog formation are effective, by far the most important cause for cloud formation is upward movement of bodies of air.

It is well known that pressure decreases with height. When a body of air rises, it moves from higher pressure to lower pressure. In so doing it must expand, and as it does so, its temperature is reduced. One can calculate the amount of cooling to be expected when the air is lifted by a fixed amount. If the air is dry and no heat is added or taken away as the air ascends, it cools at the rate of 1.0°C per 100 meters. This is known as the *dry adiabatic lapse rate*. The term "adiabatic" indicates that the parcel of air does not gain or lose heat by radiation or conduction. The term "lapse rate" connotes the rate at which temperature is lapsing, or decreasing, with altitude.

As the air rises and its temperature decreases, the relative humidity of the air increases until saturation occurs and condensation begins. The standard technique for predicting the altitude of cloud bases makes use of these principles. Have you ever noticed how uniform the heights of the bases of summer cumulus clouds are? This is an indication that the air near the surface is fairly uniform; the same amount of ascent is causing saturation and condensation.

Once a cloud has begun to form, the cooling effects caused by the expansion of the rising air are partially offset by the heat released during the condensation process. We are all familiar with the fact that evaporation causes cooling. When you step out of a bathtub or swimming pool into dry air, the cooling effects as the water evaporates are sometimes chilling indeed. When condensation occurs, the reverse is true; heat is added. The quantity of heat transferred for each gram is the same, regardless of whether there is evaporation or condensation. It is known as the *latent heat of vaporization* and is equal to about 600 calories per gram of water.

The temperature of a rising body of air in which condensation is occurring falls at the rate of about 0.6°C per 100 meters. The released latent heat accounts for a decrease of the lapse rate of 0.4°C per 100 meters from the dry adiabatic rate.

If the rate of ascent of air, which may be called the *updraft speed,* is quite high, the air may cool so fast that condensation cannot proceed fast enough to keep the air at saturation. In this case the air may become supersaturated. However, as we have already mentioned, supersaturations are not likely to exceed 1 per cent even in extreme cases.

Within a rising mass of air there are millions of condensation nuclei of the type discussed in Chapter 2. They cover a wide range of sizes; some are hygroscopic and attract water-vapor molecules, others are difficult to wet. The first detailed study of the growth of cloud droplets in such a situation was made by Wallace E. Howell at M.I.T. in 1948. He took a realistic population of nuclei and calculated how they would grow in updrafts of various sizes. This work involved a fairly complicated set of equations which took into account the size and composition of the nuclei, the changes of the concentration of the salt solutions in the droplets, the rate of heat transfer as the droplets grew, and the interactions of the droplets. He found that the large nuclei start growing first, but as the air becomes slightly supersaturated, condensation begins on the smaller ones. The equations show that once this happens the smaller droplets grow more rapidly than the large ones. The final condition is one with clouds having a narrow range of droplet sizes.

As mentioned in an earlier section, observations of cloud droplets show that their sizes extend over a fairly large range. This result is in conflict with the calculations of Howell and other investigators. However, Howell did not consider the effects of the giant nuclei, which could account for the presence of large cloud droplets. Other scientists have suggested that in order to explain the cloud-droplet populations actually observed it is necessary also to include the effects of collisions between cloud droplets.

In summary, it can be concluded that the clouds which we see day after day are indications of regions of rising air in which condensation has occurred on small condensation nuclei. The forms of the clouds depend on the character of the field of vertical motion. In the next chapter we shall examine various types of clouds and their origins.

Chapter 4

VARIOUS KINDS OF CLOUDS

Clouds come in a variety of sizes, shapes, and textures. At times the sky looks like a confused mixture of white and gray as clouds of several types move in different directions in response to the wind at their respective altitudes. When one studies the appearance of clouds, it becomes apparent that there are distinct types. This fact was recognized in 1803 by Luke Howard, in England, who proposed a classification of clouds. He followed, by one year, the famous French scientist J. B. Lamarck, who also had devised a classification scheme. Howard's system was quickly adopted and is essentially the same one used by present-day weather observers.

Howard proposed that, on the basis of the appearance of the clouds, one could group them into three fundamental classes: *stratus, cumulus,* and *cirrus*. Stratus are widespread layers or flat patches which are quite uniform. Cumulus are individual masses of clouds which develop vertically in the form of rising mounds, domes, or towers, often with tops that resemble cauliflower. Cirrus are clouds composed of ice crystals; they form at high altitudes and usually have a hairlike and silken appearance.

When a cloud has properties of two of these basic types, the composition is reflected in the name. For example, when there is a layer cloud whose surface is wavy or has the form of many cumulus clouds, it is called *stratocumulus*.

The modern classification also takes into account the

altitude of the clouds. If a layer cloud is below about 6500 feet, it is in the stratus family and is called a low cloud. If a layer cloud is at an altitude between about 6500 and 20,000 feet, it is called a middle cloud and the prefix *alto-* is added to the basic name. Thus, a stratiform cloud at an altitude of 10,000 feet is called *altostratus.* High clouds—that is, those over 20,000 feet —have the term *cirro-* as a prefix. For example, a stratiform cloud at high altitude is called *cirrostratus.*

When precipitation either in the form of snow or rain is falling from a stratus or cumulus cloud, the term *nimbus* is combined with the basic name. This yields the cloud types *nimbostratus* and *cumulonimbus.*

As was noted earlier, the separation of clouds into the various categories we have been discussing is based on the appearance of the clouds to an observer on the ground. Frank H. Ludlam, in England, has been a strong supporter of the idea that it is sensible to classify clouds according to the kind of air movements that lead to their formation and growth. From the point of view of one concerned with the processes of cloud formation, Ludlam's suggestions certainly have merit. He has proposed four basic classes: (1) Orographic clouds formed as a result of vertical motion caused by mountains or hills; (2) layer clouds formed by widespread irregular stirring of the air; (3) layer clouds formed by widespread regular ascent of air; and (4) cumuliform clouds formed by "penetrative convection," a term which will be clarified in due course.

Orographic Clouds

When air flows over a mountain range, the flow is disturbed in a way that produces a wavelike pattern such as shown in Figure 10. Clouds shaped like giant lenses are often produced in the region of rising air (see Plate II). When one examines these clouds with motion-picture techniques that speed up the action, it is found that the clouds are being formed on the upwind side and are evaporating on the downwind side. In the rising branch of the flow speeds of 200 to 2000 ft/min are likely. Although a cloud may sit over a fixed spot for hours, the lifetime of any one cloud droplet is only about ten minutes.

Mountains also play an important role in the formation of cumulus and cumulonimbus clouds. In addition to the effect the mountain has on the flow of air blowing over it, it also acts as a source of heat. The

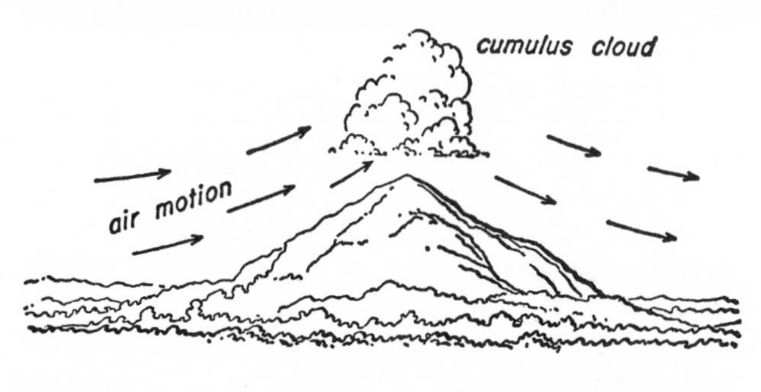

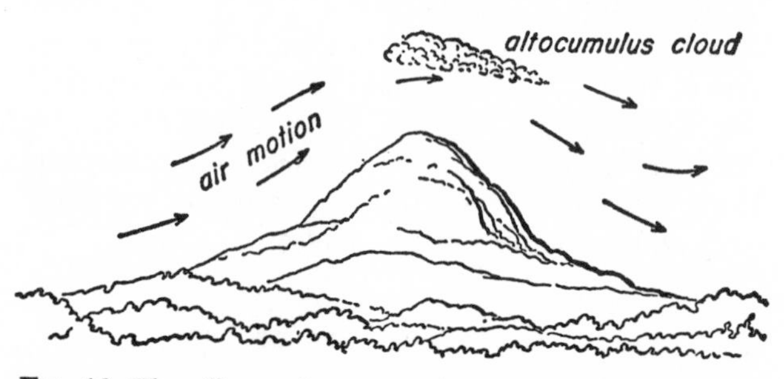

FIG. 10. The effects of a mountain on cloud formation. In the summer when the air is moist and unstable, clouds of the cumulus family often develop. When the air is stable, clouds shaped like giant lenses form as the air flows over the mountain.

sun's beating on the high ridges causes them to warm up to temperatures higher than the temperatures of the free air at the same altitudes over the valleys. As a result, once upward motion is started, the warmer, lighter air accelerates rapidly and large clouds may be formed.

Layer Clouds

In the preceding chapter we stated that as a result of cooling at the earth's surface fogs may form. When the winds are moderate or strong, stirring of the air near the surface causes the fogs to be lifted and a widespread layer of uniform cloud may be created (a stratus cloud). Variations in the pattern of the vertical motion can lead to the formation of waves or tufts and cause the cloud to become a stratocumulus (Plate III).

If you look at a weather map in the daily newspaper or on a television weather show, you almost always see regions of low pressure and so-called fronts. A *front* is the boundary separating warm air from cold air. When the two bodies of air move at different speeds, the warm air glides up the frontal surface, as shown in Figure 11. This condition leads to upward movement of air over very large areas. The ascent speeds may average only 20 to 40 ft/min, but if they continue for many hours, the air rises many thousands of feet. Far out ahead of a warm front one frequently sees cirrus clouds like those shown in Plate IV. As the front approaches, the clouds form into layers which gradually thicken as their bases get lower and lower, until rain or snow begins to fall.

Extensive cloudiness is sometimes caused by a low-pressure system even when no front is present. This happens when the winds near the ground blow in such a way that air begins to converge towards a particular region. The air, escaping compression, then starts to

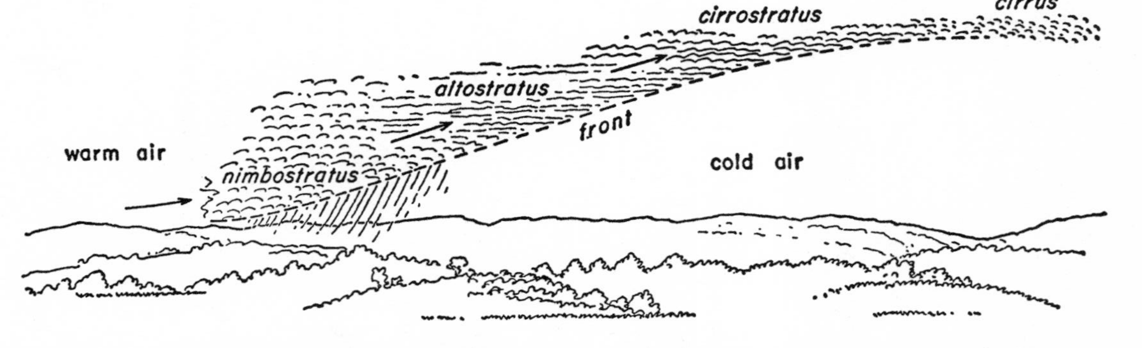

Fig. 11. Schematic drawing of a warm front, the zone separating a wedge of cold air from a body of warm air. As the air glides up the front, various types of clouds form.

rise. As in the case of warm fronts, the air ascends slowly, but it may continue doing so for a long time. Widespread layer clouds lasting many days can be produced in this way.

CUMULIFORM CLOUDS

The most spectacular clouds in the atmosphere are cumulonimbus or thunderclouds, which often produce torrential rains, hail, lightning, and thunder. They are caused by the rapid ascent of relatively small volumes of air. The term "penetrative convection" is used to convey the idea that a small volume of cloud air "penetrates" vertically through a large region of relatively undisturbed air. Clouds of this type are commonly started by warm air near the surface of the earth. Once a mass of warm air begins to rise in an unstable atmosphere it may continue the movement because it is less dense, and hence weighs less, than the surrounding air.*

In some cases only small cumulus clouds form. As the clouds grow to higher altitudes, clear, dry air mixes into the clouds and causes them to be chilled by evaporation of the cloud droplets. When this occurs, the cloud stops growing and evaporates. Small cumulus clouds normally last only about five to fifteen minutes.

On the other hand, if the atmosphere is very moist and unstable the rising air may be accelerated as it moves to high altitudes (Plate V). Updrafts exceeding 4000 ft/min can be developed. The cloud can grow into the stratosphere to altitudes over 60,000 feet. Since the lower part of the stratosphere is composed of very stable air, it acts as a deterrent to further cloud growth. It often causes the cloud tops to spread out and leads to

* For a more detailed discussion of the growth of thunderstorms see *The Nature of Violent Storms,* L. J. Battan, Science Study Series No. S 19, Doubleday & Co., Inc., 1961.

the development of "anvil" clouds (Plate VI). The lower part of a cloud such as the one shown in this plate is composed of water droplets, while the anvil cloud is made up of ice crystals. We will discuss the process of ice-crystal formation in the next chapter.

Chapter 5

ICE CRYSTALS

The growth of ice crystals in the atmosphere is one of the marvels of nature. As will be seen later, the fundamental shape of ice crystals is hexagonal, that is, six-sided, but around this theme there is an almost infinite number of variations. The variations range from flat plates with six perfectly straight and equal sides to configurations having six branches, each with an identical pattern of intricate tracery of icy fibers. Sometimes you find crystals in the form of six-sided prisms; these shapes are called needles or columns.

It is not often that you see perfect crystals falling to the ground. The atmospheres in which they form may lead to uneven growth. Collisions between crystals and partial melting may lead to irregularities. On the other hand, on very cold days with widely separated crystals falling slowly from the clouds, you sometimes have the privilege of capturing them on the sleeve of your coat and observing their exquisite beauty. In a later section are photographs of certain types of crystals—some of simple design, and one having a very complex structure. But let us, at this point, consider why water freezes, and come back later to the crystals.

Freezing of Water

If it were possible to take a small quantity of "absolutely pure" water and seal it in an "absolutely clean" bottle, you would find that when the temperature fell below the so-called "freezing point," at 32°F, the water would not freeze. As a matter of fact, if you continued to lower the temperature, it might not freeze at 22°F or even 0°F. An experiment precisely like this one has never been performed. No one has yet figured out a way to obtain absolutely pure water or absolutely clean glassware. Modern laboratory techniques allow one to obtain very clean water and glassware, but one cannot be sure that all impurities have been removed.

When one uses the most sterile techniques available, it is found that water can, in fact, be reduced in temperature–that is, supercooled–by great amounts. The amount of supercooling depends on the quantity of water involved. The smaller the quantity, the lower the temperature at which freezing occurs. Figure 12 shows a curve of the average freezing temperature for water droplets of various sizes. Experiments by many scientists all over the world, particularly those of R. G. Dorsch and P. T. Hacker, in the United States, and E. K. Bigg, in Australia, follow such a pattern. When the temperature is lowered to about –40°F, those droplets not already frozen will freeze immediately. Vincent J. Schaefer at the General Electric Laboratories in New York was the first to report, in 1946, that even the tiniest water droplets would freeze at this temperature.

There appear to be different reasons for the gradual decrease of the freezing temperature with decreasing drop diameter (shown in Fig. 12) and for the sudden freezing of droplets at about –40°F. It has been proposed that the shape of the curve indicates that the larger the volume of water, the greater the likelihood of the presence of a special particle on which the ice can begin to grow at modest degrees of supercooling. Such a particle is called an *ice-crystal nucleus*.

In the absence of an ice-crystal nucleus, ice may form only by the accidental grouping of a large number of water molecules into an aggregation resembling ice. B. J. Mason, a prominent English cloud physicist, was one of the first to investigate the details of how this happens. Let us consider the freezing process in "pure" water.

Water vapor consists of individual molecules spaced widely apart and not bound to one another—that is, some may move without influencing the others. When the water-vapor molecules condense to form a liquid, namely water, the molecules become loosely bound to one another. They may all move freely when the vessel which holds the liquid is tilted. When the liquid freezes, the water molecules become rigidly bound to one another in a pattern determined by the structure of the water-vapor molecule. The hydrogen and oxygen atoms

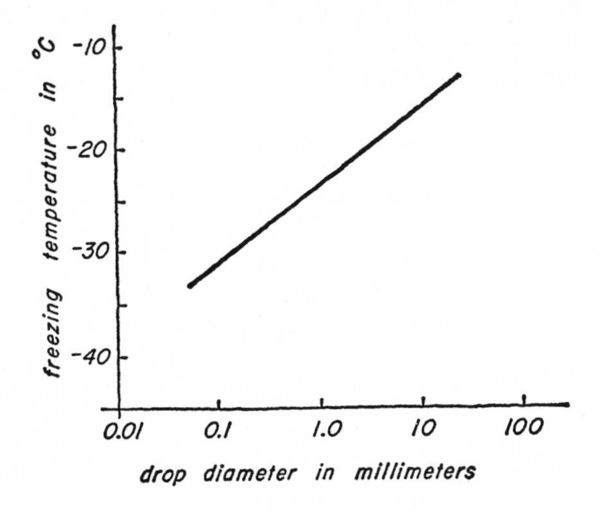

FIG. 12. An oversimplified diagram showing the variation of the temperature at which droplets of relatively clean water freeze as the size of the droplets increases. (From E. K. Bigg)

which make up H_2O arrange themselves in a tight, orderly pattern.*

The transition from liquid to solid is accomplished by a lowering of the temperature of the substance. As water becomes colder, the individual molecules composing it move more slowly, and there is a greater chance for the hydrogen and oxygen atoms to arrange themselves in the tight, orderly pattern of ice. When a sufficient number of molecules become so arranged, they act as a nucleus on which other water molecules attach themselves, and the water begins to freeze. The lower the temperature, the greater the chance of formation of an ice nucleus of this type. Calculations have shown that the likelihood increases rapidly when the temperature falls below –40°F. This explains why it is not possible to supercool clean water below this temperature. At warmer temperatures the formation of ice is usually caused by a solid particle on which the water molecules may deposit themselves.

Ice-crystal Nuclei

About fifty years ago a German scientist, A. Wegener, first proposed that certain extremely small particles in the atmosphere could lead to the formation of ice crystals. He observed such formations when the temperature fell below freezing and the humidity was sufficient to saturate the air. He called these particles *sublimation nuclei* and visualized that they might act somewhat like the condensation nuclei described in Chapter 2. Wegener suggested that water-vapor molecules would deposit themselves directly in the frozen

* See *Crystals and Crystal Growing,* by A. Holden and P. Singer, Science Study Series No. S 7, Doubleday & Co., Inc., 1960, for a more thorough discussion of the properties and growth of crystals.

state as long as the air was *saturated with respect to ice*. Let us consider what this means.

In Chapter 3 we discussed the meaning of the term "saturated." When such a condition prevails, the air over a water surface in a closed jar contains the maximum number of water molecules that it can support. If more molecules escape from the water, an equal number is forced from the air into the water. The pressure of the vapor in this case depends only on the temperature and can be calculated from well-known equations.

If the jar is suddenly cooled to a temperature of, let us say, −10°C and the water suddenly freezes, the vapor pressure of the air, for a very short instant of time, remains the same. But just at the surface of the ice the vapor pressure is less than it was at the surface of the water. The water molecules in the ice are more tightly bound than are molecules in water and cannot escape as easily as they do from water. As a result, the vapor pressure at the surface of the ice is less than it would be at a water surface of the same temperature. Because the vapor pressure of the air is higher than that of the ice, water-vapor molecules must be driven towards the ice. This migration of water molecules continues until the vapor pressure of the air equals that at the surface of the ice. When this condition is reached, the air is saturated with respect to ice.

Figure 13 shows the saturation vapor pressure with respect to water and ice for a range of temperatures. It can be seen that at some temperatures the saturation vapor pressures are greater over a water surface than over an ice surface. This fact is extremely important in explaining the formation of ice crystals and snowflakes.

Let us return now to the early discovery of the ice-crystal nuclei. Wegener inferred that they acted as sublimation nuclei when the air surrounding the particles was *saturated with respect to ice*. Since the announcement of his early research there have been some scientists who have supported his ideas. On the other hand, another group of investigators have concluded from their research that one rarely obtains ice crystals unless the atmosphere is *saturated with respect to water.* One theory proposes that crystals are formed when ice-crystal nuclei are intercepted by supercooled water droplets.

The present thinking is that a supercooled droplet will freeze if it captures the proper type of particle. However, ice crystals form on some types of nuclei if the nuclei first develop very thin films of water over their surfaces. Although the exact thickness of the film

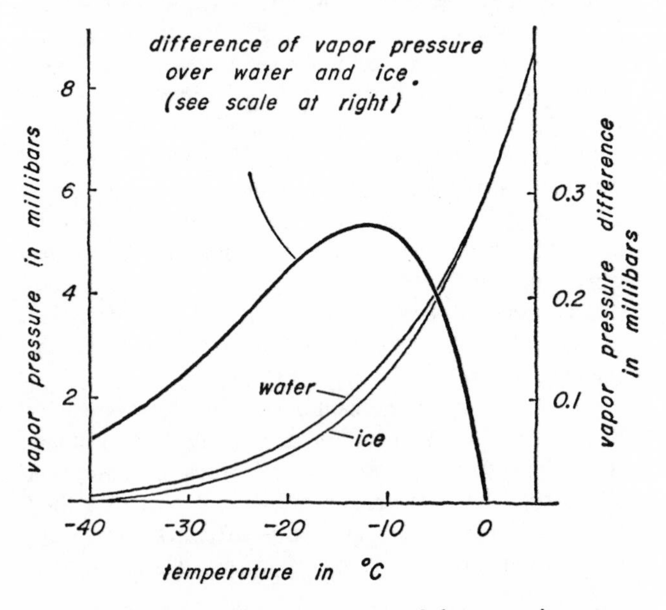

FIG. 13. Variation with temperature of the saturation vapor pressure (the vertical scale on the left) over a water and ice surface. The heavy curve shows the difference of vapor pressure over a water and ice surface. It is evident this value reaches a maximum at about –12°C.

still is uncertain, it is likely that a layer of water about five molecules thick forms over a nucleus. This process may occur even if the air is not quite saturated with respect to water. When low temperatures are encountered, this thin film may freeze and become the foundation on which other layers of ice may form.

When one talks about ice-crystal nuclei in the atmosphere, it is essential to specify the temperature involved. As the temperature is reduced, more and more particles become active nuclei. Figure 14 shows the variation of nuclei concentration at various temperatures measured in the summer in Arizona. Observations such as these have been made by many scientists in many parts of the world. They do not all agree. There are various reasons for the disagreement. First of all, there are real differences from one geographical area to the next, but also the techniques of observation differ considerably. Nevertheless, most of the observations do show that until temperatures fall to below $-15°C$, the concentration of nuclei seldom exceeds about $10^4/m^3$, and usually is much less. At lower temperatures the concentration of nuclei increases.

Sources of Ice Nuclei

There is an uncertainty about the chief sources of atmospheric ice nuclei. One important question is whether they come from the surface of the earth or from outer space.

In recent years E. G. Bowen, at the Commonwealth Scientific and Industrial Research Organization in Australia, has offered the hypothesis that cosmic dust is a major source. He has suggested that as the earth moves through regions of meteor showers, the earth sweeps up huge quantities of fine particles which settle into the atmosphere and act as ice-crystal nuclei. This theory is not based on evidence that meteoritic dust particles are good ice nuclei. Its chief supports are rainfall statistics. Bowen and others have found that when rainfall all over the earth is taken into account, there are distinct peaks which fall on particular dates, for example February 2 and March 13. In general, the rainfall peaks

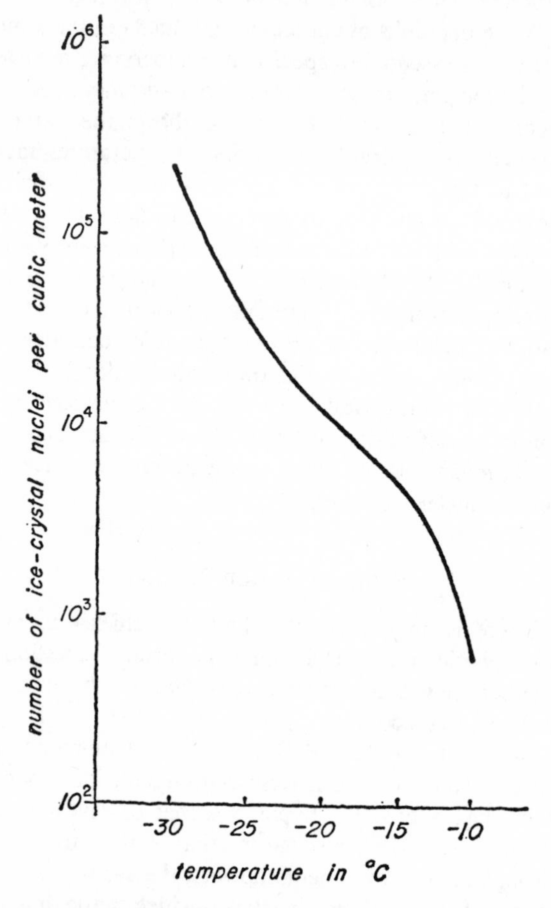

FIG. 14. The average distribution of natural ice-crystal nuclei at different temperatures, measured in southern Arizona in one summer.

occur about thirty days after the earth has passed through known meteor-dust clouds. Although Bowen does not have a great deal of support for his position, evidence of sufficient strength to reject his hypothesis has not yet been offered.

It appears that most natural ice-crystal nuclei are composed of certain types of soils. The most convincing clues are found in the fine work done by a group of Japanese scientists, particularly that of Motoi Kumai. He captured ice crystals on the stage of an electron microscope. After the crystals had melted, he found the nuclei and was able to identify most of them. In his early work, in 1951, Kumai reported that 35 out of 43 nuclei were composed of soil particles. Subsequent work has supported these results.

The effectiveness of various types of soil particles as ice-crystal nuclei has also been attacked on another front. Vincent J. Schaefer collected various types of soils and made laboratory tests. He found that certain types of clay particles are effective ice nuclei. The most effective are those called kaolinite and montorillonite. James E. McDonald, at The University of Arizona, performed a similar type of study and was able to show that wind-raised and wind-carried dusts from the deserts of the world could roughly account for the number of atmospheric nuclei which produce ice crystals at about –15°C.

Various Types of Ice Crystals

Natural ice crystals have a variety of shapes. Plate VII shows photographs which illustrate them. The most distinctive groups are the needles, plates, and so-called "dendritic crystals," those with branches.

It has been found that the crystal type does not depend on the nature or size of the nucleus. The governing factors are the humidity, and particularly the temperature, of the air in which the crystals grow. Needle crystals are most likely when the temperature is in the vicinity of –5°C. Plates are commonly formed at temperatures between –10° and –20°C, especially when the vapor pressure is not too high. Dendritic crystals grow in a moist atmosphere when the temperature is about –15°C.

As crystals fall toward the ground, they may pass through a large range of temperature and humidity conditions. As a result various types of growths may occur. Sometimes one observes needles with flat plates that have developed across the ends of the needles. Other complicated shapes have also been seen.

When ice crystals collide, they often stick to one another. They produce a snowflake, if enough of them merge. When you see large flakes of snow, you can assume that many individual crystals have aggregated. If the temperatures are sufficiently low that no melting occurs during the fall of the flakes, you can make out the individual crystals. At temperatures above 32°F the snowflakes melt and reach the ground as rain. In Chapter 6 we shall examine the natural processes of rain formation in more detail.

Chapter 6

FORMATION OF RAIN AND SNOW

In human terms the atmosphere is a source of both good and bad. When clouds scattered in a blue sky take on brilliant orange and red colors as the sun sets, the scene can be breathtaking. In a different sense, the cameo beauty of an ice crystal is equally satisfying. On the other hand, the fierce violence of a tornado or thunderstorm or a bombardment by hailstones as large as oranges will frighten almost anyone.

Probably the least exciting weather event is ordinary rain. Of course, a farmer whose crops are drying up in the fields or a city whose reservoir is emptying greets a rainstorm with joy, but the usual reaction is not so emotional. The infrequent, very beautiful or very violent phenomena of the atmosphere make a big impression, but ordinary rain and snow are like old friends. We are almost always glad to see them even though they do not bring a great deal of excitement. Also like old friends, they play extremely important roles in our lives.

Raindrops and Cloud Droplets

What is a raindrop? How does it differ from a cloud droplet? Before describing how a raindrop forms, we should know the answers to these questions.

It is common to picture a raindrop as having the shape of a pear. We know, however, that large raindrops more nearly resemble a hamburger roll (see Plate VIII). They are flat at the bottom and round at the top. As they fall, their shapes change as the water "bounces" up and down. The *surface tension* of the water makes it behave somewhat as a balloon full of water would behave at the end of a string. With very large drops the oscillations sometimes become so great that the drop breaks up into many smaller drops. When we use the term *surface tension,* we refer to a force between the water molecules which acts to pull the molecules together. You can see its effect if you slowly fill a glass with water until the water level is actually *higher* than the top of the glass. Be sure to pour very slowly and carefully.

When water droplets are small, less than a few hundred microns in diameter, surface-tension effects pull the water towards the droplet center. The result is that the droplets are virtually perfect spheres. As the drops become larger, the amount of deformation increases. At diameters of several millimeters the shape shown in Plate VIII is common.

A typical raindrop has a radius of 1 mm (which equals 1000 microns). We have noted already (Chapter 3) that an average cloud-droplet radius is about 10 microns. Let us assume for the sake of simplicity that the rain and cloud droplets are spherical. We know from geometry that the volume of a sphere is given by the formula

$$\text{Volume} = \frac{4}{3} \times \pi \times r^3$$

where π is the constant 3.14 and r is the radius of the droplet. By substituting the radii for the cloud droplet and raindrop one can calculate the volume of each. It will be found that the raindrop has a volume one million times greater than the cloud droplet. Hence, if one raindrop is to be produced, the water contained in a million cloud droplets must be gathered together.

As noted in Chapter 3, observations have shown

that the numbers of cloud droplets in a cubic centimeter of air varies from perhaps fifty to several hundred. Let us take 100/cm^3 as an average value. Raindrops are much less numerous, of course. They occur in frequencies of several hundred to several thousand per cubic meter. Let us take 500/m^3 as an average value. These figures show that cloud droplets are about 200,000 times more numerous than raindrops.

A reader with an analytical mind may wonder at this point where the water comes from to produce the raindrops. If a raindrop contains the water of a million cloud droplets but the number of cloud droplets in a cubic meter is only 200,000 times greater than the number of raindrops, there does not appear to be enough water to produce the number of raindrops actually observed. This apparent inconsistency is resolved by realizing that the region containing cloud droplets is more than ten times greater than the region containing raindrops of average sizes.

A question now faces us. How does nature bring together the water of a million cloud droplets to form a raindrop?

Condensation and Raindrops

It was once thought that raindrops were no more than large cloud droplets. Of course, it does seem reasonable to think that if condensation can cause water droplets to grow to diameters of 20 to 30 microns in ten minutes, then as a long time passes, drops of 100 to 1000 microns should grow also. The fact is, however, that raindrops do not form in this way.

For the reasons mentioned when we were discussing the formation of cloud droplets, the process of condensation does not go on indefinitely. We noted that condensation occurs when the air is supersaturated with respect to the growing droplet. A sea-salt particle in solution attracts water molecules even when the relative humidity is less than 100 per cent. This happens because we measure relative humidity relative to pure water. As already noted, the equilibrium humidity over a salt solution is lower than that over pure water. As the drop grows, the strength of the solution gets weaker and weaker. By the time droplets have grown to diameters of a few microns, the salt solution is normally so low that the droplet is relatively pure water. In this circumstance the droplets will continue growing only if the air is supersaturated with respect to pure water, that is, if the relative humidity is greater than 100 per cent. Such conditions may prevail in rapidly ascending air. However, the higher the humidity the greater are the numbers of smaller condensation nuclei which begin to grow and share the supply of available moisture.

The effects of these processes, therefore, are twofold: (1) The larger the drop the slower it grows, and (2) the higher the supersaturation the greater the number of cloud droplets. Thus it can be understood that, with the condensation nuclei normally found in the atmosphere, condensation alone cannot lead to rainfall. There is one possible exception to this rule.

When giant sea-salt nuclei are present, condensation may lead to droplets with diameters of 100 microns or so. If the clouds form near the ground, some of the drops may fall to the ground. This type of precipitation yields only small quantities of rainfall.

If condensation doesn't lead to rain, how then does rain form? Meteorologists have found that nature does it in two ways.

Ice-crystal Process for Snow and Rain Formation

In the early 1930s the famous Norwegian meteorologist Tor Bergeron proposed a theory of precipitation which, for the most part, is still regarded as valid. He recognized the importance of the fact that the saturation vapor pressures of water and ice were different at temperatures just below freezing. As you will recall from Figure 13 in Chapter 5, the saturation vapor pressure of water is higher than that of ice at the same subfreezing temperature. The difference reaches a maximum at a temperature of about –12°C. This result means that if ice crystals are introduced into a cloud of supercooled waterdrops, the ice crystals will grow while the waterdrops evaporate. Let us look at the process in steps.

Assume that originally there is a cloud of water droplets at –12°C. The air surrounding the drops would, after a short time, become saturated with respect to water. In this circumstance the droplets would be stable. They would neither grow nor evaporate, because for every evaporating molecule of water there is a molecule that condenses.

Assume now that some ice-crystal nuclei are suddenly introduced into the cloud and that a small number of ice crystals are formed. As soon as this occurs, the cloud system becomes unstable. The air is saturated with respect to water, but it is supersaturated with respect to the ice crystals. As a result, water-vapor molecules deposit on the ice crystals. As soon as this happens, the air is no longer saturated with respect to water. Consequently, some water evaporates from the cloud droplets to make up for the losses to the crystals. This evaporation again leads to supersaturation with respect to the ice, the crystal grows larger, and the cycle continues.

It is convenient to visualize the process in this step-by-step progression, but in fact the transfer of water molecules from water to ice is continuous. Ice crystals can grow very rapidly. In a matter of just a few minutes ice crystals with diameters exceeding 100 microns can grow. As long as the crystals remain in a super-cooled cloud, they continue to get larger.

Up to this point we have not considered the vertical movement of the droplets and ice crystals. As long as the droplets are small, we can safely neglect the effects of vertical motion. For example, a cloud droplet of 10-micron radius would fall at a speed of about 1 cm/sec in still air. This velocity is so low that we can assume that the droplet is carried by the air. When ice crystals several hundred microns in diameter have developed, their velocities can no longer be neglected. A flat-plate crystal 200 microns wide falls at a speed of about 20 cm/sec. The fact that the crystals fall faster than the cloud droplets is very important in the formation of precipitation.

As soon as the crystals start falling through the cloud, they begin to collide with the cloud droplets and other ice crystals of different sizes. When they collide, they stick to one another and the droplets freeze. Meteorologists call this process *coalescence.* It has been shown that when the crystals exceed about 200 microns in diameter, the rate at which they grow by coalescence exceeds the growth by deposition of molecules directly on the ice crystal. Also, the larger the aggregate of crystals the more rapid is the rate of growth by coalescence.

In winter when temperatures at the ground are below freezing, the aggregate of ice crystals reaches the ground in the form of snow. When temperatures at the ground are above 32°F the snow melts and we get rain.

Sometimes when the temperature at the ground is below freezing, there may be a warm layer of air above the ground (Fig. 15). For example, the temperature near the ground may be 28°F while the temperature at 4000 feet is 38°F. As the snowflakes pass through the layer where the temperature exceeds 32°F they melt and become raindrops. Then, as they fall into the subfreezing surface layer of air, they freeze again and reach the ground as *sleet,* small frozen pellets of ice.

If the layer of cold air near the ground is not deep enough or cold enough for the raindrops to freeze, they reach the surface as supercooled water. When they strike objects on the ground, the water freezes very quickly. When this happens, we call it *freezing rain.* Plate IX shows the effects of such an event. Trees, transmission lines, houses, etc., become coated with a layer of ice. The effects can create wonderlands of beauty with glistening ice hanging from anything exposed to the freezing rain. Unfortunately, as is often the case, not only does every cloud have a silver lining

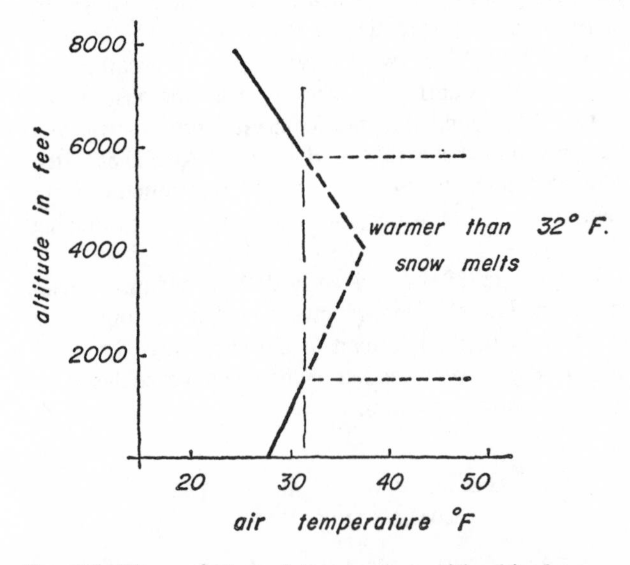

FIG. 15. The variation of temperature with altitude on a day when freezing rain can form. As snow particles fall through the zone above 32°F (bounded by broken horizontal lines) they melt. When they fall into the region near the earth they are cooled to below 32°F, become supercooled, and freeze on striking cold objects.

but many silvery objects have tarnished edges. Freezing rain can cause tremendous damage. Wires sometimes snap from the weight of the ice. Tree branches often break. And, last but not least, the effects of icy roads on the accident rate are appalling indeed.

It is clear that the ice-crystal process can account for much of the precipitation that falls, particularly during the winter months and in the colder latitudes. The capture of a snowflake composed of many perfectly symmetrical crystals is clear evidence of its efficacy.

During the late 1930s and early 1940s it was felt that the ice-crystal process accounted for virtually all the precipitation that reaches the ground. However, there were some skeptics, particularly G. C. Simpson in England. When World War II took weathermen from many countries to the tropical regions of the world, they found that the ice-crystal process was not the only one that could lead to rain. Airplanes often flew over clouds whose temperatures were much warmer than freezing but which, nevertheless, were producing rain.

In the late 1940s and early 1950s evidence began to accumulate showing conclusively that in convective clouds rain is often formed in the absence of ice crystals. This discovery has led to the so-called coalescence theory of rain formation.

Formation of Rain by the Coalescence Process

Cumulus clouds over tropical oceans usually produce raindrops in the absence of ice crystals. One can fly over the tops of clouds from the instant they begin to form until they advance to the rain stage. In the Caribbean area they typically appear first at an altitude of about 2000 feet. Their tops grow at speeds of about

400 ft/min, and by the time they have reached 10,000 feet, they often contain raindrops whose diameters may be 500 microns. To one who has spent most of his life in higher latitudes it is surprising to see clouds only 8000 feet thick producing showers, but it is a sight often seen in the tropics. The temperature at the summits of such clouds is about 7°C. Unquestionably, in clouds such as these the raindrops are produced in an all-water process.

An obvious question arises. How do such raindrops form? This leads to another question. Does the same thing happen outside the tropics?

We have noted already that if all the droplets in a cloud are small and of uniform size, the cloud is a stable system. All the droplets fall very slowly and at the same speed. As a result, the number of droplet collisions is small; it would take a very long time before a million droplets could merge.

On the other hand, if for some reason there were introduced into the cloud some drops whose sizes were greater than the cloud droplets, the situation would be changed drastically. A droplet of 10-micron radius falls at a speed of 1 cm/sec, while droplets of 50-micron radius fall at a speed of 26 cm/sec. These larger drops, falling faster than the cloud droplets, would overtake them and collide with them. Assuming the simplest type of motion, one can calculate readily the number of collisions. As shown in Figure 16, a large falling drop sweeps out a vertical cylinder whose cross-sectional area is equal to that of the drop. In a minute the drop will have fallen a distance which is easily computed if one knows the drop's speed of fall. For a drop of 50-micron radius the volume swept out in one minute would be about 0.1 cm^3. If the concentration of 10-micron cloud droplets were $100/cm^3$, the number of collisions would be 10, that is 0.1 $cm^3 \times 100/cm^3$. In fact, the number would be greater, because as droplets collide and coalesce with the falling drop, the latter increases in size and the cross section of the "cylinder" increases. More properly, instead of a cylinder we should use a slightly tapered cone. However, let us neglect this effect since it merely complicates the description.

If all the colliding droplets were to merge with the falling drop, the rate at which the large drop grows would be easy to calculate. However, it is found that all collisions do not lead to coalescence. You may be astonished to learn that two drops of water bounce off one another, but they do. Next time you are swimming, splash the water with your hand or foot and watch the little droplets of water closely. You will see some of them bounce off the surface one or more times before they finally fall back into the lake or swimming pool.

The fact that all collisions do not lead to coalescence is most dramatically shown in high-speed movies of col-

FIG. 16. An oversimplified picture of how a large, falling drop sweeps up smaller cloud droplets.

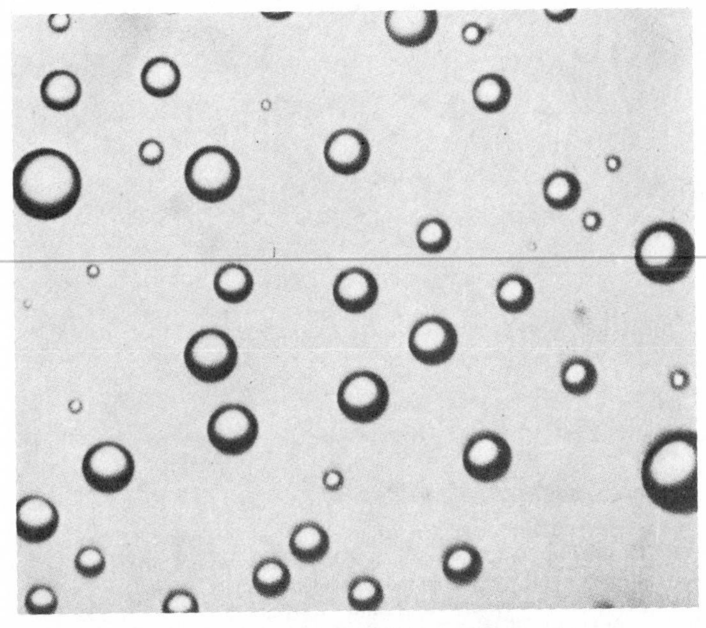

PLATE I (above). Photograph of cloud droplets through a microscope. The largest droplets are about 40 microns in diameter.

PLATE II (below). Altocumulus lenticularis clouds (that is, altocumulus clouds that look like giant lenses) formed by airflow over a mountain. The clouds in the center of the photograph are composed predominantly of water droplets. *(Courtesy U. S. Department of Commerce, Weather Bureau)*

PLATE III (above). Stratocumulus cloud. A layer cloud with regular lines of convective motion. Clouds like this one are found at low altitudes and are composed of water droplets. *(Courtesy U. S. Department of Commerce, Weather Bureau)*

PLATE IV (below). Cirrus clouds. They occur at high altitudes and are composed of ice crystals. Ice crystals form at a particular altitude, and as they fall, they are carried horizontally by the wind and produce cloud tufts which have the shape of hooks. *(Courtesy U. S. Department of Commerce, Weather Bureau)*

PLATE V (above). Rapidly building cumulus clouds. Continued growth of these clouds leads to the formation of cumulonimbus clouds and rain.

PLATE VI (below). Cumulonimbus cloud with a well developed anvil top. The anvil clouds are composed of ice crystals while the lower parts of the cloud are made up of water droplets. This cloud is a typical thundercloud that produces rain, lightning, thunder and sometimes hail.

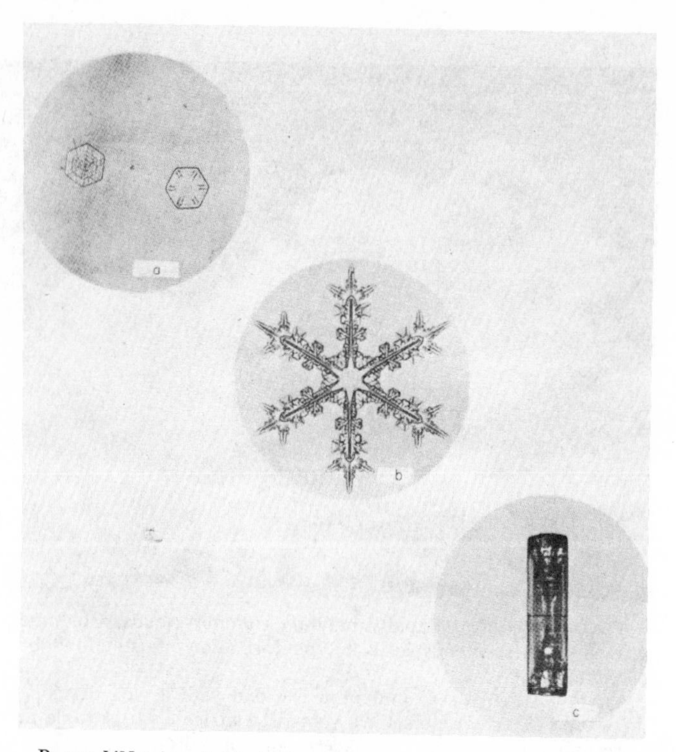

PLATE VII (above). Various types of ice crystals: (a) flat plate; (b) stellar crystal; (c) column.

PLATE VIII (below). Photograph of a waterdrop falling in still air. This drop had a horizontal diameter of 6.5 mm and was falling at 8.9 m/sec. *(From Choji Magono)*

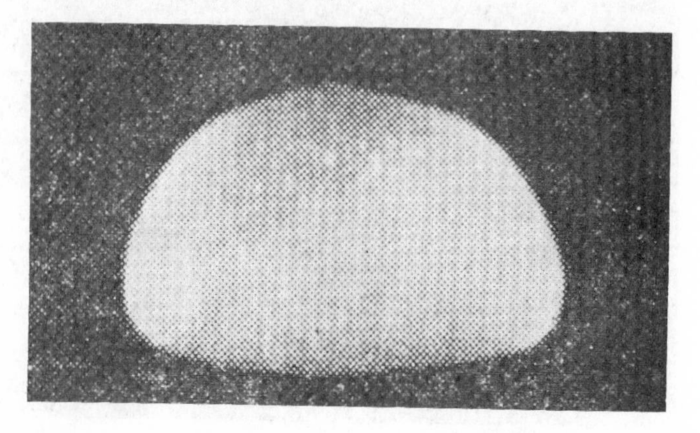

PLATE IX (above). Freezing rain causes branches of trees to become coated with ice and the weight sometimes is enough to produce the effects seen in this photograph. *(Courtesy U. S. Department of Commerce, Weather Bureau, and New York Power and Light Co.)*

PLATE X (below). Formation and growth of a rain echo in a cumulus cloud over southwestern Ohio. The numbers under each panel give the time in minutes and seconds. The top of the panel represents an altitude of about 20,000 ft; the lower part is the ground.

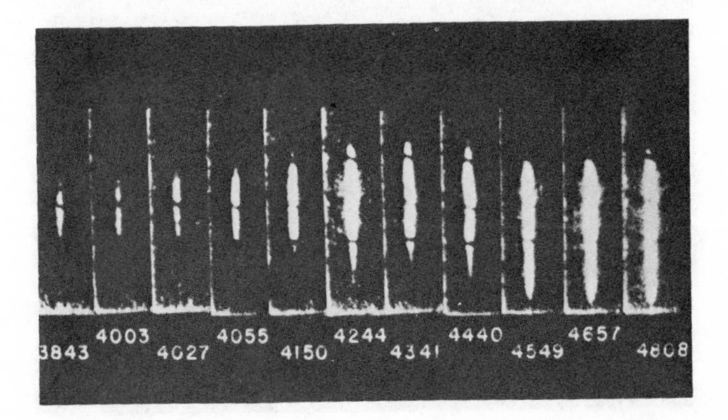

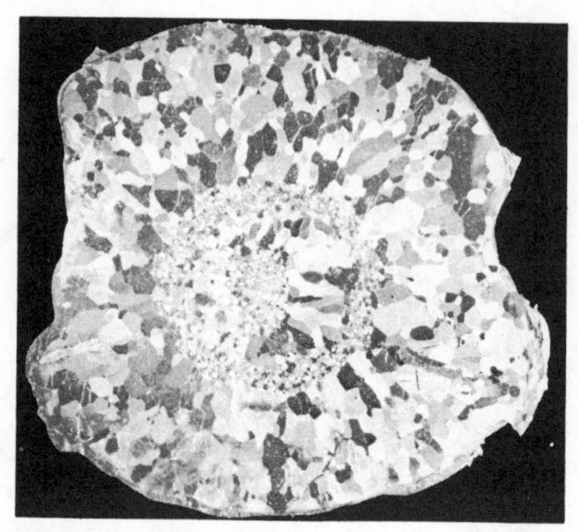

PLATE XI (above). Thin slice cut from a hailstone about 4.5 cm in diameter. The stone was photographed by passing polarized light through the ice. This procedure permits the identification of the individual ice crystals of which the stone was composed. The regions of small crystals appear as milky ice under ordinary light while the regions of large crystals are clear ice. *(Courtesy Vincent J. Schaefer)*

PLATE XII (below). The appearance of the top of a deck of supercooled stratus cloud about 20 minutes after dry ice pellets were dropped into it by an airplane which flew a race track pattern. It can be seen that the dry ice modified the cloud. It caused the formation of ice crystals which fell out of the cloud leaving a "hole." *(Courtesy Vincent J. Schaefer)*

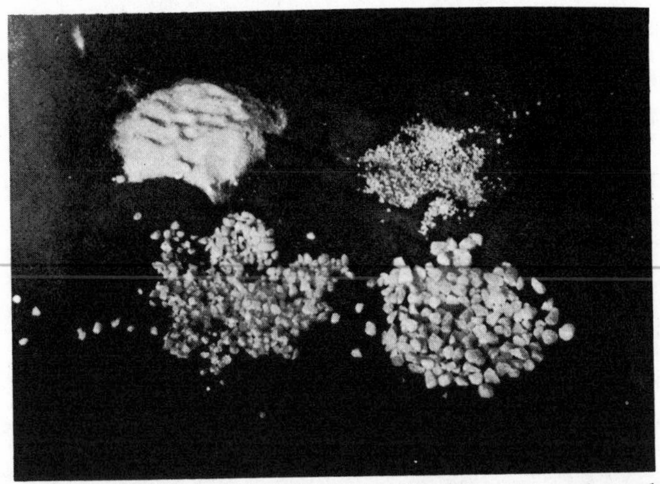

PLATE XIII (above). Dry ice pellets produced by an ice crusher and used on several cloud-seeding projects. The largest pieces are about ½-inch in diameter; the smallest are like sand particles. With simple adjustments of crusher larger or smaller pellets can be produced.

PLATE XIV (below). An early model of the U. S. Forest Service, Project Skyfire, ground-based silver-iodide generator. The tank on the right holds propane gas which produces a flame into which a silver-iodide-in-acetone solution from the gallon bottle on the left is injected. The two metal cylinders are slipped over the flame in order to allow proper burning of the flame and protect it from wind and weather. *(Courtesy U. S. Forest Service)*

PLATE XV (above). Airborne silver-iodide generator of the U. S. Forest Service, Project Skyfire. A silver-iodide-in-acetone solution is stored in the upper tank and is fed to the burning compartment just below the tank. Spark plug ignites acetone sprayed into the burner. Once lighted, it continues burning as long as fuel is sprayed.

PLATE XVI (below). A B-17 airplane used by the University of Chicago to seed warm cumulus with a water spray. The water was dumped through a large valve in a 400-gallon tank in the bomb bay. When the water fell into the slip stream, it broke up into small drops and looked as shown in this photograph.

liding waterdrops. At speeds of 7000 frames per minute, the film shows that smaller drops will sometimes sink into larger drops and come bouncing out, reminding you of a gymnast on a trampoline. The small drop deforms the surface of the larger one without breaking it. It still is not entirely clear why some droplets behave this way while others merely merge with the larger drops. It has been proposed that a very thin layer of air between the two water surfaces prevents them from merging. Laboratory tests have shown that if the drops are allowed to collide in the presence of strong electric fields, they almost always coalesce. This condition has led some scientists to believe that electrical forces may play an important role in the precipitation process.

At any rate, it should be clear that if the sweep of droplets proceeded as in Figure 16, and if the fraction of collisions leading to coalescence were known, it would be an easy matter to calculate the growth rate

FIG. 17. A more realistic picture of the motion of small cloud droplets as a large drop falls through a cloud. Some of the small droplets move with the air as it flows around the large drop. As a result, they fail to collide with the large drop. The smaller the cloud droplets, the smaller the chances that they will follow a collision course.

of a drop whose initial radius was 50 microns. Unfortunately, the situation is not quite so simple as illustrated in Figure 16. While the large drop falls, the air moves around it and carries some of the smaller drops along. Figure 17 presents a more realistic picture of the normal state of affairs. Of the cloud droplets in the projected cylinder, some collide with the large drop, others do not. The fraction of the droplets within the cylinder that actually hit the large drop is called the *collision efficiency*. It depends on various factors, including the size of the drops as well as the properties of air. Figure 18 presents a curve showing the collision efficiency for a drop of 100-micron radius in clouds of various-sized droplets. When the cloud droplets are very small, most of them move around the large drop. As the droplets increase in size, greater fractions of them collide with the large drop. There is a whole series of curves such as the one in Figure 18 for drops of different sizes.

If one knows the collision efficiency and the fraction of the collisions leading to coalescence, the rate of growth of the large drop can be calculated. Once drops of 100 microns in diameter are produced in a cumulus cloud, they can grow to form raindrops 1 millimeter in diameter in a matter of 15 to 20 minutes (if the cloud is made up of many large cloud droplets). In clouds with very small cloud droplets growth by coalescence is slower.

But where, you may ask, do the large drops come from in the first place? This is an important question.

In Chapter 2 we discussed giant sea-salt nuclei. These particles of salt, whose radii in the dry state are larger than a few microns, may have a part in production of the sparse large drops, which lead to precipitation. This idea, first proposed by Frank H. Ludlam, in England, recognizes that giant salt nuclei can grow to radii of 30 to 50 microns by condensation alone.

Furthermore, their concentrations may be on the order of 100 to $1000/m^3$, a value similar to the concentration of raindrops.

Other scientists have suggested other ways to start the coalescence process. It has been proposed that elec-

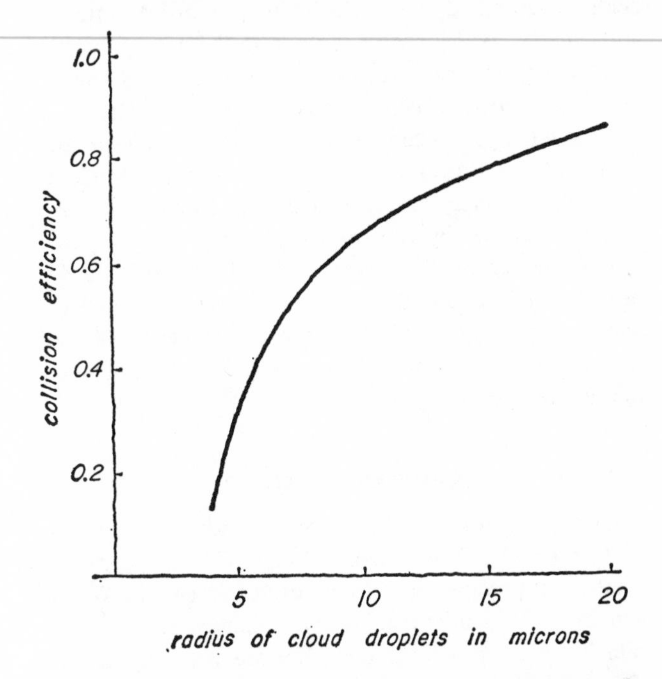

FIG. 18. Calculations show that when a drop falls through a cloud, the collision efficiency depends on the radii of the larger drop and of the cloud droplets. Collision efficiency is the ratio of the number of droplets colliding with a large drop to the total number swept out as the drop falls (those within the cylinder shown in Fig. 16). When the efficiency is 1.0, all the droplets in the cylinder strike the large drop. The curve in this illustration shows the variation of collision efficiency of a drop 100 microns in radius as it falls through a cloud of droplets of radii given on the horizontal scale. (From Irving Langmuir)

trical attraction may lead to collision and coalescence of very small droplets and the production of droplets with radii 30 to 50 microns. If this were true, it would not be necessary to have such giant sea-salt nuclei. The importance of these electrical effects, still in doubt, is being investigated, particularly by Bernard Vonnegut and Charles B. Moore of Arthur D. Little, Inc., who are leading proponents of the idea that electrical effects are very important in the growth of precipitation.

In summary, we can say that although some of the details of the process are still not well understood, it has been definitely established that in tropical cumuli, rain is produced by the so-called *coalescence process.*

It seems reasonable to expect that if the coalescence process is effective in tropical clouds, it should be effective in similar types of clouds in other regions of the world. That this is true has been established only in the last ten years.

Radar Observations

Until meteorologists began to use radar sets for observing clouds, it was almost impossible to know at what point precipitation drops formed in clouds. When you observe clouds visually, you cannot be sure that rain has been produced until you see it falling out of the cloud. You must realize, however, that the raindrops did in fact form five to ten minutes earlier. In the time required for the drops to fall from the central part of the cloud to its base, the cloud top could have grown 5000 to 10,000 feet. So it is difficult to know the properties of the cloud *at the time* when the raindrops *first* formed.

Radar helped to solve this problem. Most of the radar sets commonly used by meteorologists do not "see" cloud droplets, but they do detect precipitation particles.* Thus, if an observer examines a building cloud and sees the initial radar echo from that cloud, he can tell that the echo represents the region containing the first precipitation drops.

In various parts of the world radar has been employed to observe the altitude of the initial echo. Plate X shows a series of radar observations of the evolution of a precipitation echo in a cloud which formed over southwestern Ohio. Each panel represents a vertical cross section through the echo. The horizontal scale of each panel is about 12 miles, while the vertical scale extends from the ground to 20,000 feet. Under each photograph is the time in minutes and seconds. These observations taken at short time intervals show that the initial echo appeared in the altitude range from 7000 to 12,000 feet. The temperature of the initial echo top was about 6°C. The region of raindrops grew rapidly until the bottom of the echo (and the rain) reached the ground.

Radar observations such as these show clearly that even in regions far from tropical oceans precipitation may be formed by coalescence rather than by the ice-crystal process.

Systematic studies have been made of the relationship between the vertical extent of the clouds and the likelihood of precipitation. We have mentioned that over tropical oceans, cloud bases are usually at an altitude of about 2000 feet. It is found that in the central United States the bases of convective clouds are often at an altitude of about 5000 feet and that active clouds grow at a rate of 500 to 1000 feet per minute. Unlike their tropical counterparts, these clouds seldom produce raindrops until their tops penetrate the 15,000-foot level, where the temperature is below 0°C. For exam-

* See *Radar Observes the Weather,* by L. J. Battan, Science Study Series No. S 24, Doubleday & Co., Inc., New York, 1962.

ple, observations show that only 20 per cent of the clouds reaching the level of 18,000 feet contain precipitation particles.

In the semiarid regions of the United States, such as New Mexico and Arizona, even larger clouds are needed to produce rainfall. In these areas the cloud bases usually are at 10,000 to 12,000 feet, and only 20 per cent of the clouds penetrating to the 25,000-foot level produce rain.

Figure 19 shows a schematic drawing of the heights and thicknesses of the clouds that yield radar echoes in 20 per cent of the cases. It is obvious that clouds are higher over the central United States and still higher over southwestern United States than those found over tropical oceans. Also, if precipitation is to occur, increasingly taller clouds are needed in less humid regions.

On the basis of the kind of information contained in Figure 19, it has been argued that the coalescence process frequently initiates precipitation in convective clouds in all three regions. In the central United States the air is drier than over the tropical oceans; a greater lift is needed before a cloud is formed. Furthermore, with a higher (and colder) cloud base, the cloud top must be higher in order to condense the same quantities of liquid water. Thus, clouds in the central United States that produce rain are greater than in the tropics. In arid regions ever greater lifts are needed to produce liquid water concentrations like those found in tropical cumuli.

At the present time it is the opinion of most cloud physicists that if the bases of convective clouds are at warm temperatures, the coalescence process may account for the formation of precipitation. On the other hand, when the cloud base has a low temperature, ice crystals may form in the cloud when it is only a few thousand feet thick. Once crystals have appeared, they can grow to diameters of several hundred microns in a

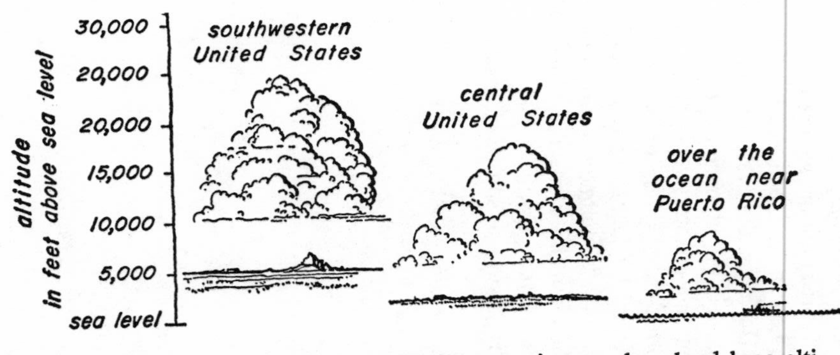

FIG. 19. Schematic drawing of the differences in cumulus cloud base altitudes and thicknesses in three different geographical regions. Observations have shown that the chance of precipitation particles is about 20 out of 100 in all three clouds.

matter of minutes. After that their growth by collision with other crystals and cloud droplets will rapidly produce particles of precipitation size. It is known that once the growing particles have reached diameters of a few hundred microns, be the particles water or ice, the coalescence process is the only one that counts.

The processes of precipitation in stratiform clouds have not been studied in as much detail as those in convective clouds. However, the evidence indicates that the ice-crystal process is the dominant one.

Looking back on this chapter we see that we have been concerned with the formation of precipitation of normal sizes—that is, with waterdrops whose diameters are no greater than a few millimeters, or with snowflakes that would be about the same size when melted. Now we shall take up the growth of hail, a form of precipitation that is always a source of wonder and often a peril.

Chapter 7

HAIL

Hail has fascinated people for centuries. When a thunderstorm begins to pepper the ground with hail, most people stop to look, pick up the hailstones, and examine them. In most parts of the world hail particles are usually small—about the size of green peas. In certain unfortunate areas, however, it is not uncommon to find them as large as walnuts. Once in a while they get to be as large as tennis balls. These are, of course, hailstones in their most destructive form.

When storms of large hail form, they can wipe out a farmer's crops in minutes. A beautiful field of corn can be laid bare so that only the mangled stalks remain. Fruit trees which have taken years to reach maturity can be stripped to the bark. Damage to all types of agricultural products runs into millions of dollars every year.

Farmers aren't the only ones who suffer from these icy bombardments.

The windows broken by hailstones run into the thousands every year. Tile roofs have been battered and shattered. Automobiles have been pockmarked beyond repair when caught in a shower of large hailstones. And, last but not least, airplanes have been badly damaged. An encounter is bad enough when planes are on the ground; in flight it could be catastrophic.

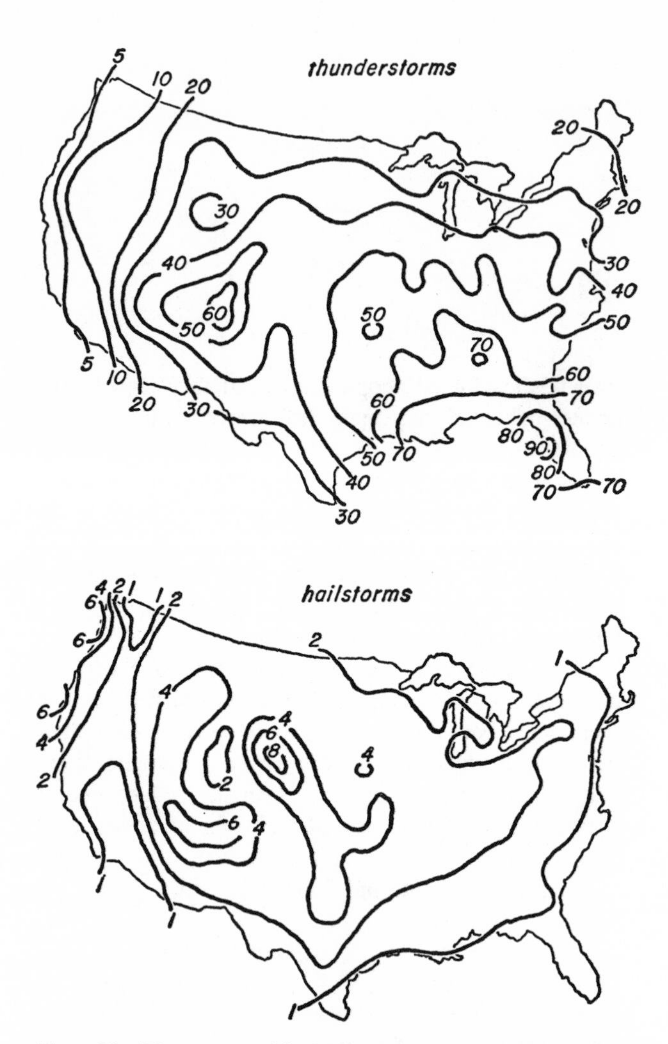

Fig. 20. The geographical distribution of thunderstorms and hailstorms in the United States. The numbers represent the average number of days with storms per year. Central Florida has the maximum number of thunderstorms, about 90 per year; the maximum number of hailstorms occurs in eastern Colorado, about 8 per year at any one location. (From U. S. Weather Bureau data)

Where Does Hail Form?

Hail has been observed at one time or another in almost every place where thunderstorms occur. But in some areas it is very seldom seen, while in others it is quite common.

The average number of thunderstorms and hailstorms each year over the United States is shown in Figure 20. The state of Florida experiences more thunderstorms than any other state, but the occasions of hail at the ground are few. Over the Great Plains of the United States the frequency of thunderstorms is less than over Florida, but the number of hailstorms is much greater.

Other regions of the world also have unusually high occurrences of damaging hailstorms. Northern Italy and the Caucasus region of Russia are among them.

As already noted, the size of hailstones varies from one storm to the next. Figure 21, taken from a report by W. Boynton Beckwith, of United Air Lines, shows the relative frequency of hailstones found in storms in the vicinity of Denver, Colorado. In this area large stones are common. On the average, seven reports of stones over an inch in diameter are collected in this area every year.

Several important questions have interested meteorologists for quite a number of years. Why do some thunderstorms produce hail while others do not? Why are certain particular regions, such as the Great Plains, favorite spots for hailstorm formation? We still have no satisfactory answers, but scientists all over the world are giving these problems concentrated attention and a clearer understanding is emerging.

Not too many years ago most ideas on this subject were little more than speculation. Now we are beginning to learn more about the conditions that accompany the formation of hailstorms. Detailed studies of thunderstorms with radar equipment and flight measurements are leading to an improved understanding of hail-producing thunderstorms. Also, new techniques for studying the hailstones after they reach the ground are yielding information about how they are formed.

Some things are known about conditions leading to hail-producing thunderstorms over the United States. The atmosphere must be unstable; that is, it must have temperature and moisture conditions that permit the

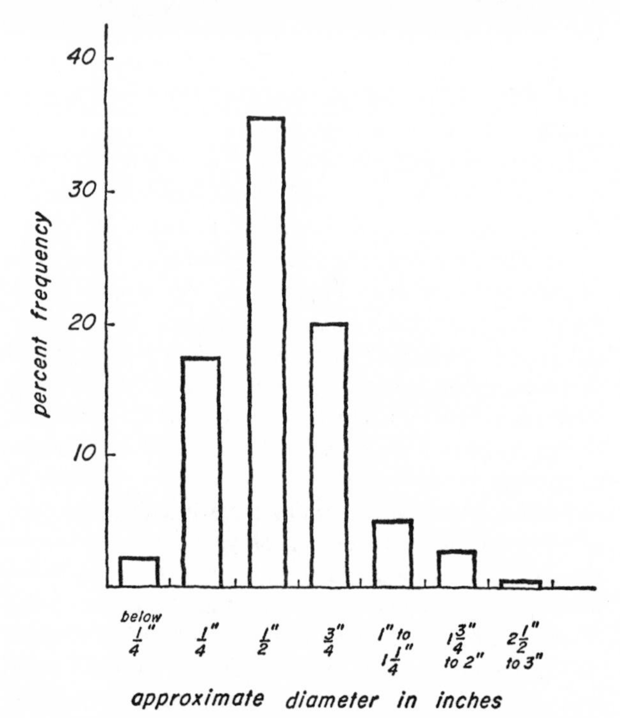

FIG. 21. The per cent occurrence of the largest hailstones in eastern Colorado. For example, 20 per cent of the storms have stones with diameters between three quarters and one inch. (From W. Boynton Beckwith)

development of thunderstorms with strong updrafts.* When there is a stream of warm, moist air blowing toward the north near the earth's surface and a jet of drier air blowing from the west at higher elevations, conditions are ripe for severe thunderstorms. These conditions often exist over the Great Plains. Moist air pours northward from the Gulf of Mexico. Dry air speeds towards the east from over the mountains.

If some kind of atmospheric disturbance causes a body of unstable air to be lifted, the unstable properties of the atmosphere take over. One may think of this in terms of the analogy of a delicately balanced boulder on the side of a mountain. It might remain in the same place for a long time if undisturbed. But if someone gives it a little push, it starts rolling down the mountain at ever-increasing velocities until it reaches the valley floor.

In an unstable atmosphere a volume of air displaced upward continues accelerating upward. Sometimes the acceleration goes on until the volume reaches the base of the stratosphere, at altitudes over 45,000 feet. In this way large and violent thunderstorms may be formed. They produce intense rainfall and lightning, and sometimes they produce hail. It is not clear why it is only sometimes. Some investigators have proposed that a critical feature is the height of the freezing level above the cloud base. In Florida, for example, the cloud bases are about 2000 feet above sea level, while the freezing level is at about 16,000 feet, and the thunderstorms seldom produce hail. On the other hand, in eastern Colorado the cloud bases are often about 9000 feet above sea level, while the freezing level is at 14,000 feet. Thunderstorms with these features often produce hail. Herbert Appleman, of the U. S. Air Force, has sug-

* See *The Nature of Violent Storms,* L. J. Battan, Science Study Series No. S 19, Doubleday & Co., Inc., 1961.

gested that with a thick layer of warm cloud, rain is produced before much hail can begin to form.

In the last few years there has been an increasing interest in the role played by upper-level winds. This interest has been stimulated by the experiences of weather forecasters, who have noted that a jet of high winds is often present when severe thunderstorms occur. Frank H. Ludlam, in England, has proposed an hypothesis similar to one advanced earlier by Joe Fulks, of the U. S. Weather Bureau. It visualizes an arrangement of updrafts and downdrafts that would permit a growing ice particle to remain in a thunderstorm long enough to grow to a large size. Also, the proposed pattern of vertical motions is such that the updraft can accelerate to unusually high velocities. It has been proposed that in the upper parts of such clouds the updraft speeds may exceed 100 m/sec (224 mph).

Quite evidently, much more research is needed to resolve some conflicting views on hail-producing thunderstorms. At this point it is clear that in order to produce large hailstones there must be large thunderstorms with very strong updrafts. Also, the chances of large hailstones are increased when the wind increases with height and causes a shearing of the cloud in the direction of the strong winds.

Some Properties of Hailstones

A normal property of hailstones is that the ice of which they are composed is not uniform. Some parts of almost every stone are composed of clear ice while other parts are composed of milky ice. The opaque appearance is caused by trapped air bubbles. In large stones one sometimes finds alternate layers of clear and opaque ice.

Anyone who has had occasion to see hailstones knows that they frequently are nonspherical. Many stones are shaped like a pear. Since they fall with the apex upward, the freezing water accumulates on the nearly flat bottom surface.

Details of the interior sections of hailstones have been studied by slicing thin sections and examining them under various types of light. This technique, first developed in Switzerland by M. de Quervain and Roland List, has been very revealing. Plate XI shows a photograph of a thin section of hail viewed under polarized light. The hailstone is composed of many hundreds of ice crystals. Each crystal has a flat face which makes a particular angle with the ray of light passing through the thin slice of ice. The color of the light which would be seen depends on that angle. A large crystal appears as a large area of uniform color, a small crystal as a small area of uniform color. The region of opaque ice is made up of small crystals and trapped air bubbles, while the region of clear ice is composed of large crystals. The center of the hailstone shown in Plate XI was made up of many small crystals and air bubbles. It was surrounded by a layer of large crystals (clear ice), then a thin layer of small crystals (opaque ice) and finally a thick layer of clear ice.

Why should there be alternate layers of large and small crystals? The explanation is found in the rate at which the water is collected and frozen. When the stone falls through a region of low cloud water and intercepts small quantities of supercooled water, that water may freeze almost instantly. On the other hand, if the falling stone accumulates large quantities of water, it cannot freeze instantly. Instead, the stone gets wet, and the freezing proceeds slowly as large crystals grow and force out the air in the water.

In order to better understand the internal structure of the hailstone let us consider the manner in which a stone grows.

Growth of a Hailstone

In attempting to reconstruct the series of events which lead to the formation of a hailstone, various items of information must be taken into account. First of all, we know something about the stones themselves, their sizes, and their crystal structure. Any theory on hail formation must explain their growth to diameters of more than 5 cm. Also, it is necessary to explain the layers of clear and opaque ice.

Furthermore, it is known that thunderstorms which produce hail are accompanied by tall cumulonimbus clouds with strong updrafts. The quantity of supercooled cloud water must be high to allow for rapid growth of the stone by coalescence. This quantity, known as the *liquid water content,* usually has magnitudes between 0.01 and 1.0 gm/m^3, but in thunderstorms it may reach peaks of 4 to 5 gm/m^3. In theory, it is possible to obtain as much as about 8 gm/m^3, but such large values have never been measured with reliable instruments.

Another important feature of a thunderstorm is the character of the updraft. A new theory of hail formation, proposed by Frank H. Ludlam, requires that it be tilted (Fig. 22). It also requires that the updraft be fairly steady over a period of thirty minutes to an hour.

It is easy to show that when large hailstones, say 3 cm in diameter, are involved, the number per cubic meter of cloud air which can be grown in any one thunderstorm is small. In Chapter 6 we noted that an average cloud droplet has a diameter of 20 microns, while an average raindrop has a diameter of 2 mm. Since the volume of the drops is proportional to the cube of the diameter, the water of a million cloud droplets must be combined to form one raindrop. A hailstone 3 cm in diameter has a volume about 10 billion

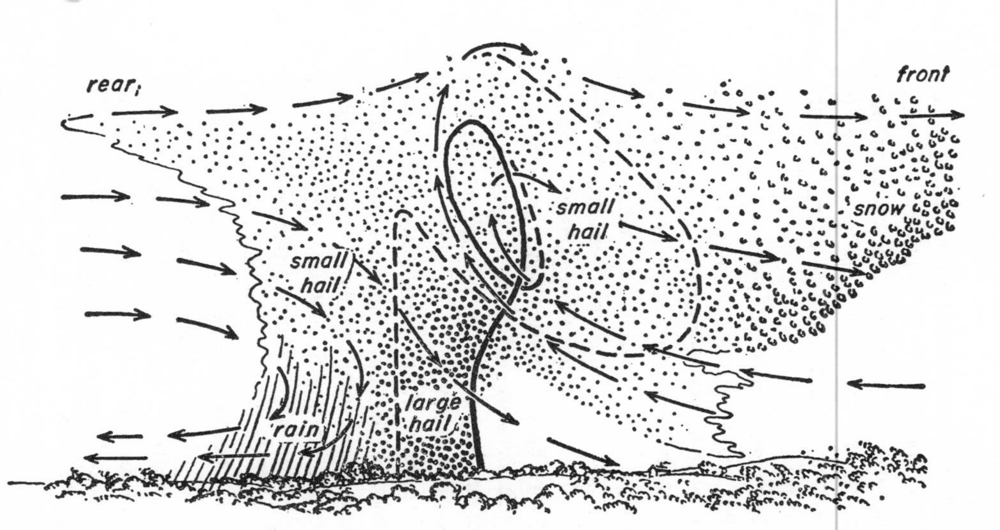

Fig. 22. Frank H. Ludlam's proposed model of a severe hailstorm. The arrows show the flow of air through the cloud. On the front side the air blows into the cloud at low elevations and out of the cloud at the higher elevations. The dashed line shows the path followed by small hailstones. They form in the upper parts of the cloud and move through the air in the manner shown. Near the rear portion of the cloud they melt to form rain. In some cases small hailstones are carried back to the upper parts of the cloud by the air currents and form large hailstones by sweeping up supercooled water. This is shown by the dashed line which becomes a solid heavy line.

times larger than a cloud droplet. In other words, about 10^{10} droplets must be accumulated to produce one 3-cm hailstone.

We also mentioned earlier that a typical cloud contains about 100 droplets per cubic centimeter. Thus, in order to produce one 3-cm hailstone the cloud droplets in 10^8 cubic centimeters of cloud must be gathered together. This amounts to one hailstone every 100 cubic meters of cloud air.

These simple calculations show why very large hailstones are rather scarce. An exceedingly large amount of cloud water must be accumulated.

Let us return now to the first stage of the hailstone growth. The "hail embryo," as it is sometimes called, probably is a large waterdrop which has frozen while being carried by the updraft to cold regions of the cloud. Some investigators have proposed that the embryo is composed of ice crystals which have agglomerated. Microscopic examinations of the hailstones have not yet given conclusive answers.

At any rate, once a frozen particle about 1 mm in diameter is present in the supercooled region of a thunderstorm, it can grow very rapidly by collision with supercooled drops. At one time it was proposed that the intercepted supercooled droplets froze on impact to form opaque ice. This type of growth would continue until the stone fell below the freezing level and the ice began to melt to form a water layer. It was suggested that a rapid increase of the strength of the updraft would then carry the stone back up to the cold regions of the cloud, where the water covering the stone would freeze to form clear ice. A second layer of opaque ice would then form before the stone again fell below the freezing level. The process of ascent and descent across the freezing level would continue to produce layers of clear and opaque ice as the stone increased in size. Certain aspects of this theory have been rejected. It would be too much of a coincidence to expect a stone to rise and fall above and below the freezing level. Also, the layers of clear ice actually observed are too thick to have been formed by the freezing of a thin film of water. Furthermore, it is possible to explain the layers of clear and opaque ice without requiring that the stone fall to levels where the ice could melt.

A second theory is that a hailstone may be formed when an ice particle falls directly from the cloud top to the ground. The layers of clear ice are explained by noting that if excessive quantities of water are accumulated, the stone gets wet and freezing proceeds slowly. It has been found that water contains air dissolved in it. When the water freezes slowly, the air is forced out of the ice. On the other hand, when freezing is rapid, the dissolved air forms little bubbles which are trapped inside the ice.

At first glance it may be astonishing to hear that a layer of water with a temperature of –10°C freezes slowly. Remember that it is necessary to remove the so-called "heat of fusion," which is given off when water freezes. For every gram of water that freezes, about 80 calories of heat are released. These calories serve to warm the ice and the water. For the freezing to continue, the heat must be carried away from the stone by the air. Exactly the reverse process is in effect in the case of melting ice. Ice cubes in a glass of water melt slowly because 80 calories must be introduced for every gram of ice which melts. Since the necessary heat is supplied slowly, the ice cubes melt slowly.

When large quantities of supercooled water are swept along by the falling stone, it is not possible to carry away the released heat fast enough for all the water to freeze immediately. The temperature at the surface of the hailstone increases, freezing proceeds slowly, and clear ice is formed. Thus we can explain the formation of clear ice in a hailstone without having this ice descend to warm temperatures. It occurs when the stone falls through a region of the cloud where the liquid-water content is high.

When the liquid-water content of the cloud is low, the falling stone accumulates relatively small quantities of supercooled water. In this case freezing can proceed rapidly, and air bubbles cannot be forced out of the ice. When the number and size of the colliding cloud droplets are very small, the droplets can freeze almost on impact. As air bubbles are trapped, the ice becomes opaque.

The process involving a single traverse through the cloud by a growing ice particle accounts for some of the smaller hailstones, those less than about 1 cm in diameter. But if the growth of the very large stones, those greater than 3 to 4 cm in diameter, is to be explained, one traverse through the cloud will not allow the stone to sweep out enough water. It must make several up and down trips. It is not necessary, however, as was once assumed, that the stone rise and fall through the freezing level, but it is essential that the stone remain in the supercooled region of the cloud.

The Ludlam Theory

As noted earlier, Frank H. Ludlam has attempted to construct a theory of hail formation which takes into account the known facts and which meets the known requirements. Figure 22 shows a simplified sketch of his conception of a hail-producing thunderstorm. The cloud has formed in an atmosphere in which the wind increases with height. As a result, the updraft tilts in the direction of the wind. One important feature of this cloud is that the air motions associated with the rain and the cold air flowing out of the bottom of the cloud cause the cloud updraft to be constantly in the state of generation. The air at the upper level of the updraft is being carried away. In this cloud the updraft changes little over a period of thirty minutes.

A hailstone forms in the supercooled region of the cloud and begins to grow by coalescence. As it grows, its diameter increases, and the rate at which it falls through the cloud air increases. But the strength of the updraft is strong enough so that the stone still ascends relative to the ground (Fig. 23). As long as the stone falls through the cloud air, it grows by coalescence.

By the time the updraft has carried the stone to the top of the cloud, the wind has carried it to a weak part of the updraft, and the stone may fall out of the updraft and begin descending towards the ground. Then,

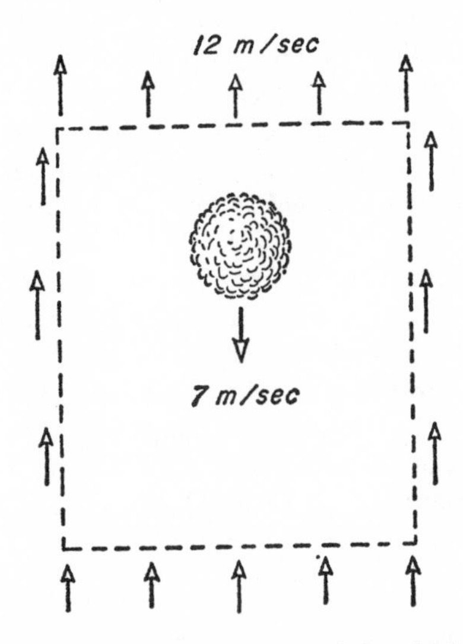

FIG. 23. How a hailstone may be carried to higher altitudes even when it is falling relative to the air. Consider the dashed lines to enclose a region of air moving upward at a speed of 12 m/sec. If the air inside the region were still, the stone would fall through it at its so-called terminal speed, let us say 7 m/sec. In this circumstance the hailstone would be moving away from the ground at a speed of 5 m/sec.

after descending for many thousands of feet, it again falls into the main part of the updraft and begins its second trip up through the supercooled part of the cloud. A series of such trips can lead to very large hail. An ice particle starting with a diameter of 1 mm can grow to 3 cm after about thirty minutes of exposure to the high liquid-water contents in the powerful updraft.

At the present time Ludlam's theory is still under intensive study. It appears to explain the known facts. Unfortunately, the number of good observations of hailstorms is quite small. For example, in order to make a satisfactory evaluation of the theory it is desirable to be able to observe the hailstones in the cloud. Just a few years ago this seemed to be a hopeless aim. However, in the last few years new radar techniques have been developed which bring it within the realm of possibility. By means of three radar sets and careful observation procedures it is possible to obtain good estimates of the size and number of hailstones in the cloud.

With attention increasing all over the world, we can be optimistic that in the near future our understanding of hailstorms will increase to the point where they can be more accurately predicted. It may even become possible to develop means for controlling them. There have been extensive efforts in this direction, but so far they have not been very fruitful. We shall look further into this aspect of weather control in Chapter 10.

Chapter 8

SCIENTIFIC BASIS OF CLOUD MODIFICATIONS

You can be quite sure that the first cave man who came home to find his cave full of rain water wished he could do something to change the weather. Many primitive tribes in remote parts of the world had, and some continue to have, rain gods to whom they pray for water. Rain dances among the American Indians have been passed on from generation to generation. At the annual powwow of thousands of Indians at Flagstaff, Arizona, various tribes demonstrate their colorful rain dances. Sometimes it rains even before they finish.

Today we scoff at the practices of the aborigines. We all know that pounding a drum and dancing around a blazing fire is not going to work.

In more recent history many people who were prepared to dismiss incantations to strange gods have believed that the weather could be changed by noise and explosions of gunpowder. During the Napoleonic Wars it was noticed that after large battles it often rained. It was argued that cannon and musket firings were the cause. Subsequent investigations have convinced almost everyone that the interpretation of the rain following the battle was not as many had presumed. Yes, the rain often did follow the battles. However, the interpretation of this result becomes easier if you were to say that the battle was fought before the rain. In the days of Napoleon, troops and guns were transported by horses and wagons. It was necessary that the ground be dry in order to move swiftly. As a result, the generals planned their attacks for periods of dry weather. As we all know, however, the normal state of affairs outside desert regions is that dry periods may last for a few days, possibly a few weeks, but sooner or later it rains again. Thus, we are led to conclude that the reason why wet days often followed battles was that nature was taking its normal course.

An important lesson to be learned from this account is that the relation between the weather and human activity is not as obvious as it sometimes seems.

The weather is always changing. To quote a well-known saying, "The most normal thing about the weather is its abnormality!"

We are accustomed to speak of the "average rainfall." Many do not understand the meaning of this term. For example, after fifty days with zero rainfall and fifty days with 0.5 inch, the average rainfall would be 0.25 inch. Thus the average represents a quantity that never occurred. In fact, in any locality the quantity of rainfall usually varies from zero to perhaps several inches in one day. This large variability makes it difficult to evaluate tests to increase rainfall. We shall return to this point in the next chapter.

Although we should recognize the difficulties in evaluation, this should not be, and has not been, a deterrent in scientific efforts to devise means to modify clouds and weather.

The first major step in the direction of weather modification came in 1946. At that time Vincent J. Schaefer and Irving Langmuir at the General Electric laboratories in Schenectady, New York, performed a series of experiments which have had tremendous repercussions all over the world. They demonstrated clearly and without doubt that certain types of clouds could be modified. A few handfuls of dry ice dropped from an airplane into the appropriate clouds quickly caused the infected parts of the cloud to disappear while snow crystals slowly drifted towards the ground. These experiments were the foundation upon which cloud physicists have built what is sometimes called "the scientific basis of weather modification."

The Use of Dry Ice for Modifying Supercooled Clouds

In order to plan intelligently and execute experiments to modify anything, including clouds, it is essential that you start with a fair knowledge of how nature does it. Once this has been established, the next step sometimes is exceedingly simple—at least, it seems simple after someone has succeeded in doing it. This certainly is true of the work of Schaefer and Langmuir. Let us examine the facts and the steps they took in making their epochal discoveries.

Many clouds are composed of water droplets which are small, fairly uniform in size, and supercooled. Such cloud systems are stable and may last for many hours or even days. As we noted in Chapter 6, one way to upset the equilibrium would be to introduce ice crystals. Schaefer and Langmuir, along with many other scientists, were aware of these facts, but they did not stop here. They asked themselves, why not introduce some ice crystals before nature does and see what happens.

Thus the next step in the research was to find a good method to produce many ice crystals. The first successful one made use of dry ice, that is, solid carbon dioxide. It can be easily shown in a home freezer that if a small piece of dry ice is dropped into a supercooled water cloud, it leads to the formation of millions of tiny ice crystals. The experiment is worth trying.

If you have a home freezer, all you need in the way of apparatus is a black cloth, a flashlight and, of course, a few pieces of dry ice. First, drape the cloth inside the freezer on the floor of the freezing compartment and up one side. Then flash the light so that you can look across the beam towards the black cloth. Now breathe into the box. You will see a grayish cloud of tiny water droplets which, in many respects, is similar to a natural cloud. The movement of the air in the freezer causes the cloud particles to move up and down, but they retain their dull gray appearance.

Now drop a piece of dry ice into the box through or near the light beam. Keep your eyes on the box now because important things are about to happen. In the cloud of droplets there will begin to appear little sparkling centers of light. The number will increase rapidly, and they will sink to the bottom of the box. You probably have surmised already what is happening. The sparkling particles are ice crystals. The explanation for the bright sparks of light is simple: the ice crystals have flat faces and the light is reflected off them as off the face of a mirror. The spherical water droplets, on the other hand, scatter the light in all directions. The crystals fall to the bottom of the box because they grow quickly by the ice-crystal process to sizes at which they fall rapidly (Chapter 6).

For some years there was some doubt about the exact process by which dry ice formed ice crystals. It now is known that it lowers the temperature to those levels at which there can be formations of ice crystals directly from the water vapor. The number of ice crystals formed by a pellet of CO_2 depends on the air temperature and moisture and on the pellet size. For example, a pellet 1 cm in diameter will produce about 10^{11} nuclei at $-10°C$ before it completely evaporates.

Once Schaefer and Langmuir had performed these tests in the laboratory, they took to the atmosphere. In the vicinity of Schenectady they found a widespread, uniform layer of supercooled stratus cloud. The pilot of the small airplane they were riding was told to fly a pattern resembling a race track. Small pellets of dry ice were dropped along the track. The airplane then flew around and around the same track as the scientists waited expectantly. In a matter of just a few minutes they saw that the texture of the cloud was changing to that of an ice-crystal cloud. With great excitement they watched while the series of events which had occurred in the laboratory freezer was repeated on a huge scale before their eyes. By the time twenty minutes had elapsed, they saw that they had unmistakably caused great changes in the cloud. As shown in Plate XII, the dry ice had "cut a hole" in the cloud. This experiment was repeated many times. When the airplane was flown down into the hole, it was found that ice crystals were floating towards the ground. Some of the crystals sometimes reached the ground in small quantities in the form of snow.

When Schaefer and Langmuir first reported their findings, they were confronted on the one side by skepticism, on the other by great enthusiasm. In the latter category was a small group of people who immediately jumped to the conclusion that if you could cause small quantities of snow to fall from a thin cloud by dropping CO_2, you could cause a great quantity of snow to fall from a thick cloud by the same method. Years have passed since 1946, and this extrapolation still has not been proved. In the meantime, there are still some meteorologists who accept it. We shall return to this matter when we discuss the subject of rain making. Let us continue at this point an examination of the scientific basis of cloud modification.

Silver-iodide Particles as Ice-crystal Nuclei

After it was clearly demonstrated that supercooled stratus could be altered by dry ice, scientists began to search for other substances that could produce ice crystals at temperatures close to freezing. In Chapter 5 we mentioned various natural substances, such as clay particles, which lead to ice crystals. However, they are not at all effective until temperatures fall below about –10° to –15°C. For the purposes of modifying clouds you need substances effective at higher temperatures.

Bernard Vonnegut, who was working with Schaefer and Langmuir at General Electric, was the first to discover the most effective substance. He was well aware that an ice crystal has a hexagonal structure of known dimensions. He reviewed tables giving the crystal structure of hundreds of chemicals. This search revealed that silver iodide had crystal properties very similar to those of ice.

In the tradition of his fellow workers, he made many tests in the laboratory freezer. The first step was to produce tiny crystals of silver iodide. After trying many schemes, Vonnegut found that if silver-iodide powder was exposed to high temperatures, it would vaporize. As it cooled, it formed minute particles of silver iodide whose diameters were of the order of 0.01 to 0.1 micron. When these particles, in the form of a fine smoke, were blown into a supercooled water cloud, they transformed it to an ice-crystal cloud in much the same fashion as did dry-ice particles.

The silver-iodide particles acted like seeds on which the ice crystals grew. The term "cloud seeding," so common today, comes from such an effect.

There still is some question of the precise mechanism by which a silver-iodide particle acts to form an ice crystal. At one time it was thought that the ice crystal was formed by the direct deposition of water molecules in the ice form. At the present time there is strong evidence that an extremely thin film of water molecules (less than five molecules thick) forms on the silver iodide. The water arranges itself into an ice-like structure and freezes. Subsequent growth is by the direct growth of the ice phase. This process is being actively studied at the present time.

Vonnegut and others have shown that silver-iodide particles produce ice crystals when the cloud temperatures fall below about –5°C. They also have found that lead-iodide crystals are effective at about the same temperature, but this chemical is not as easy to handle as is silver iodide.

Some of the advantages of silver iodide over dry ice are that it can be easily stored and can be dispersed either from the ground or from the air, provided that the air currents carry the particles into the clouds. For these reasons, silver iodide has been used extensively in cloud-seeding tests in all parts of the world.

Techniques for Dispensing Dry Ice and Silver Iodide

Seeding of supercooled clouds can be a simple matter. All that is required is that dry ice be broken into small pellets and dropped into a cloud from an airplane. This can easily be done with an ice crusher mounted on the airplane. A typical one grinds up the ice to produce pieces ranging in size from a millimeter to a centimeter. Plate XIII shows a photograph of the output of one kind of dry-ice dispenser. The crushed dry ice was passed through a series of screens to separate various sizes. The largest pieces are about 1 cm in diameter, while the smallest particles are like sand.

The quantity of dry ice needed to modify a cloud

depends on the thickness of the cloud and the purpose of the seeding. In general, it runs in the vicinity of a few pounds per mile of flight of the airplane.

Another technique for seeding with dry ice has been used by the Russians. A cage is built on the side of an airplane and pieces of dry ice are placed in it. As the airplane flies through the supercooled clouds, the air is cooled and ice crystals are produced. Note that in order for this technique to be effective it is necessary that the airplane fly through the cloud. When pellets of dry ice are dropped, the airplane can fly over the top of the cloud.

Dry ice has an advantage over other seeding agents in that it produces ice crystals at all temperatures below 0°C. However, a serious disadvantage is that it has to be put directly into the cloud one wishes to modify. Also, once it has been dropped, it evaporates and is gone forever. Silver-iodide crystals, on the other hand, are so small that they move with the air. They can linger in the air for a long time, ready to form ice crystals when a suitable cloud comes along–except if sunlight has already caused the silver-iodide particles to lose their effectiveness. The rate of decay of nuclei depends, to a certain extent, on the technique used to produce them. But, in general, the number of ice-producing particles decreases by a factor of 10 for every hour of exposure to sunlight. For example, let us say that a certain generator dispenses 10^{13} particles into a cubic meter of air. The number will diminish to 10^{12} when exposed to sunlight for an hour. Therefore, one generator discharging nuclei directly into the clouds puts out as many active nuclei as 10 generators whose nuclei are in sunshine for an hour before they enter the clouds.

There is a large variety of silver-iodide generators. Most of them involve the burning of a solution of silver iodide in the highly inflammable liquid, acetone. It should be noted that silver iodide is toxic and must be handled with care. Before going on, we should point out that powdered silver iodide does not dissolve in pure acetone. However, it does dissolve in a solution of potassium iodide* in acetone. A potassium-iodide solution is prepared first, then silver iodide is added. The concentrations employed in practice have varied from 5 to 20 parts of silver iodide in 100 parts of the potassium-iodide-in-acetone solution.

Once the silver-iodide-in-acetone solution (which we shall call "the solution") is mixed, it can be burned in various ways. For example, one can impregnate charcoal or coke with the solution. When these substances are burned, they lead to the formation of tiny silver-iodide crystals.

A popular design of ground-located generators makes use of a propane gas flame into which the solution is injected with a hypodermic needle. Since acetone is highly inflammable, it burns readily. The silver iodide is vaporized in the flame. When the fumes cool off, silver-iodide crystals of the order of 0.01 to 0.1 micron in radius are formed by condensation. Plate XIV shows a silver-iodide generator of the propane type. It consumes solution at a rate of about one quart an hour.

Seeding with ground-based generators has one serious drawback. It is sometimes very difficult to be sure that the silver-iodide particles produced at the ground are carried to altitudes where the temperatures are below –5°C and reach these altitudes over the target area (Chapter 9). This is particularly important when one is seeding in the summertime. It is common over most parts of the United States to find the –5°C temperature at a level between 17,000 and 20,000 feet. Thus, when silver-iodide generators are at the ground,

* Sodium iodide may also be used.

the particles must ascend many thousands of feet before they can reach the parts of the clouds where they may have some effect. The particles do not rise straight up at a high speed. Instead, they are carried horizontally by the winds and lifted by updrafts and slow, turbulent air motions. Also, it must be recalled again that the longer it takes for the particles to reach the sensitive parts of the clouds, the greater the possibilities that sunlight will take its toll.

One situation where seeding from the ground is not plagued by these difficulties is in some mountainous regions in winter. In some places the clouds may shroud the peaks. If the air temperatures on the mountaintops are below –5°C, silver-iodide generators at the summits will certainly lead to the conversion of supercooled water droplets into ice crystals.

Airborne Generators of Silver Iodide

The surest way to place silver-iodide particles where and when they are wanted is to use generators mounted on airplanes. For many years the only satisfactory generators available were those designed by Australian scientists working with E. G. Bowen at the Commonwealth Scientific and Industrial Research Organization. In more recent years the U. S. Forest Service has developed a good airborne generator.

One of the reasons for the delay in the development of a suitable airborne device has been the fact that acetone is highly inflammable. Also, a solution of silver iodide in acetone is corrosive and toxic. It is common practice to use stainless steel in the construction of equipment which will come into contact with the solution. Research has developed techniques for overcoming the difficulties involved.

The new Forest Service equipment, shown in Plate XV, has a number of unique features which are worth pointing out. As in the case of the Australian burner, it does not use an auxiliary fuel such as propane. Instead, the acetone itself is the fuel. Inside the tail of the burner the acetone is caused to form a spray of fine drops. A spark plug ignites the spray, and the burner acts something like a jet engine. Air flows through the front ends, mixes with the acetone, and the mixture burns with a flame that shoots from six to twelve inches out the tail. When the vaporized silver iodide leaves the flame, it condenses to form billions of crystals.

Figure 24 shows the number of crystals produced by one type of generator for each gram of silver iodide consumed. It is clear that the lower the temperature, the greater the number of effective nuclei. As was noted in Chapter 5, this is also true with natural nuclei. If cloud seeding is to be practical, it is necessary to increase the number of nuclei effective at temperatures above –15°C. The illustration shows that at –10°C the number of nuclei is about 10^{13} per gram.

An airborne generator such as the one shown in Plate XV consumes a 10 per cent solution of silver iodide in acetone at a rate of about 4 gallons per hour. As a result, the seeding rate is about 36 grams of silver iodide per mile of flight when the airplane is flying 90 miles per hour. One can calculate that at the output of the generator the concentration of nuclei is about $10^{13}/m^3$. Since it has been estimated that about 10^3 to $10^4/m^3$ are needed for rainfall stimulation, the concentrations at the generator are about 10^9 to 10^{10} too high. To overcome this difficulty it is necessary to fly upwind from the target area so that the particles may spread out into a larger volume. After the nuclei become dispersed through a cylindrical volume two kilometers in diameter, the nuclei concentration is reduced from 10^{13} to about $4 \times 10^4/m^3$, a value close to the quantity considered to be appropriate. It appears that periods of the order of one to two hours are required for this to occur, but this point still is uncertain.

The generator shown in Plate XV was very cleverly designed by Donald M. Fuquay and H. J. Wells, of the Forest Service, to assure maximum safety and convenience. The entire system is outside the airplane, suspended from a bomb rack. The tank holding the acetone solution is the upper section, with the burning chamber just below it. There is no pump in the system.

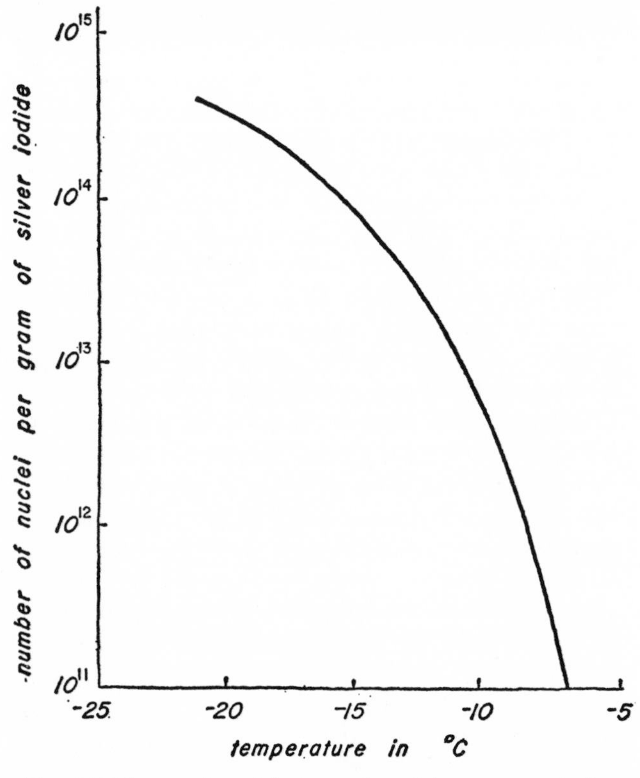

FIG. 24. The number of ice-crystal nuclei per gram of silver iodide produced by the ground-based silver-iodide generator designed by the U. S. Forest Service, Project Skyfire.

As the airplane flies through the air, pressure is built up at the inlet to the tube marked with the arrow. This increased pressure is impressed on the top of the fluid, forcing it into the burning chamber. The flow of acetone is turned on and off by a valve in the pressure line.

The seeding generators described in this chapter are only a few of the many that have been used. There is no question that they are capable of producing ice crystals in supercooled clouds. There are still many questions about the effects of the ice crystals on rainfall, hail, and other weather phenomena. We shall return to this subject in the next two chapters.

MODIFICATION OF "WARM" CLOUDS

If dry ice or silver iodide is to be effective, it is necessary that the clouds be supercooled. Is there any way to modify clouds whose temperatures are warmer than freezing?

In Chapter 6 we mentioned that natural rainfall may be produced in two ways: the ice-crystal process and the coalescence process. Dry ice and silver iodide are the seeding agents which have been used in clouds in which the ice-crystal process normally is active. The coalescence process is the one which produces rain in warm clouds. We noted earlier that this process is effective when there are some cloud particles about 100 microns in diameter. This fact gives us a hint of how to modify warm clouds. Add some large cloud droplets. Two schemes have been employed for this purpose.

If clouds were seeded with giant salt nuclei, the particles could grow to large sizes by condensation. Tests of this technique have been made in various parts of the world. Fournier d'Albe carried out a series of experiments in Pakistan in which salt particles were dusted into the air. It was reckoned that the updrafts would carry the particles into the clouds. In several

African countries salt particles were shot into clouds with rockets. The aim of these tests was to increase rainfall, but the results were inconclusive. In some cases, though, they were encouraging.

A second way to modify warm clouds was proposed in 1948 by Irving Langmuir. He suggested that one could stimulate the coalescence process by adding water droplets of diameters of 50 to 150 microns. Several groups in the United States and Australia carried out seeding trials in which water was sprayed into convective clouds. Plate XVI shows water being dumped from a 400-gallon tank in the bomb bay of a B-17 airplane in one of a series of experiments by scientists at The University of Chicago. The water, pouring through a valve having an area of about 100 square inches, was atomized as it hit the airstream. Other investigators have equipped aircraft with spray bars. The rates of seeding with water have ranged from 10 to 450 gallons per mile of flight. E. G. Bowen, in Australia, reported some success in his tests. The research group at The University of Chicago carried out a series of tests which led to the conclusion that water seeding could initiate rain. Details of these tests are discussed in the next chapter.

Chapter 9

ARTIFICIAL STIMULATION OF RAINFALL

When the news spread that Schaefer and Langmuir had modified clouds by dropping dry ice into them, the potential benefits were quickly recognized. All over the world the need of water was rising. Many places already were suffering from severe shortages. Farmers, businessmen, and politicians immediately began to see the tremendous practical advantages of a reliable means to increase rainfall. When silver iodide was discovered, the optimism ran even higher. Here was a scheme which did not even require an airplane. The growing expectations were fed by the extravagant claims of a few meteorologists who set up rain-making companies.

Commercial Cloud Seeding

In the early 1950s some commercial cloud-seeding firms were stating that rainfall could be increased by 50 to 100 per cent or more, depending on local conditions. The grounds for such allegations were weak indeed, but farmers with crops parching in the fields were willing to reach for straws. Cities forced to ration water were prepared to take the necessary risk.

It is not difficult to understand why many people were willing to gamble their money on cloud-seeding projects. If you had a corn crop worth $100,000 withering from lack of rain and if an investment of $5000 might conceivably save it, would you take the chance? In such a case it is not only the cost of the cloud seeding that counts, but also the so-called "benefit-to-cost ratio." If the financial considerations were coupled with the assurance of a weatherman that cloud seeding had a chance of success, the possible benefit of saving the crop would probably override the fear of losing an additional $5000.

During the late 1940s and early 1950s we knew almost nothing about the effects of cloud seeding on rainfall at the ground. True, the tests by scientists at General Electric had caused some light snow to fall from thin cloud layers. But the quantity of precipitation was very small and for the most part unmeasurable.

When the results became known, most scientists regarded them with interest but uncertainty. The attitude was, "Show me more—convince me." This skepticism certainly must be regarded as the proper view. When a scientist presents a new finding or a new theory, it is his responsibility to support it with convincing evidence.

On the other hand, some scientists were not so cautious. The results of the early tests were immediately extended. It was reasoned apparently that if small quantities of snow could be made to fall from a thin cloud, large quantities could be made to fall from a thick one. On the basis of this *assumption* the modern era of commercial rain making began.

Some of the commercial cloud-seeding attempts appeared to be very successful. In places plagued by prolonged periods of dry weather, the start of cloud seeding was accompanied by the start of the rains. Such a series of events can be, and has been, misinterpreted.

People begin to worry about rainfall when it does not come at the right time and the right place. Rarely will a community or a business become concerned about rainfall unless there has been a deficiency. When the quantity of rainfall fails to be "normal," this fact is well known. Newspaper, radio, and television weather reports keep a running account of such things. During a particularly dry spell the number of successive days without rain is often reported. As the number of dry days grows, concern grows. Before long some people begin to ask, "Why can't we do something about this?"

In some instances a commercial cloud-seeding firm has entered the scene at this point and proceeded about the business of dispersing silver iodide. If it rains, everyone is happy and the rain maker is suitably rewarded. But before being too quick with congratulations we should ask, are we sure the silver iodide caused the rain?

One aspect of the weather that is well known but often forgotten is the fact that it is very changeable. Cycles of warm and cold periods, wet and dry periods, cloudy and clear periods are the rule rather than the exception. The cycles occur over periods of weeks, years, and centuries. The rainfall records have never shown a place where the rainfall suddenly stopped and never fell again. If a dry spell occurs and no rain falls day after day, you can be sure that this drought cannot go on forever. It may conceivably continue until the grass in the fields has been burned brown, or the city restricts baths to two a week, but sooner or later it will rain again.

Recognizing these facts, we can see why some commercial seeding operations have apparently been successful. They came into being after prolonged spells of dry weather. The dry cycles were followed by periods of above-normal rainfall. The seeding which was taking place concurrently may have been completely incidental. Nature was working in its normal manner.

At this point a question will almost surely occur to you. Hasn't it been proved that cloud seeding can increase rainfall?

Unfortunately, the answer is no. At the present time there is some evidence that in certain situations it may be possible to cause small but important increases in rainfall. But it has not been proved beyond a reasonable doubt that cloud seeding can cause rainfall.

Let us first examine the reasons for such an inconclusive answer and then return to the evidence for optimism.

Statistical Aspects

The term "statistics" means many things to many people. To some it is a means of misleading others.* You often hear, "You can prove anything with statistics." On the other hand, it has been said, "You can't prove anything with statistics."

In fact, when the rules of statistics are understood and properly applied, statistics neither lie nor mislead. For the record, they do not prove anything strictly. But statistics do represent the only satisfactory means for *disproving* something. Before being shocked by this statement, stop and think what it means.

Let us say that you have a coin and suspect that it is not balanced in such a way that when it is flipped, a head would be just as likely to appear as a tail. For convenience let us call a perfectly balanced coin an "honest" one. If you were to flip your coin a hundred times and the result were 50 heads and 50 tails, you could rightly conclude that the coin was honest. On the other hand, let us say that the 100 flips gave 55 heads and 45 tails. Such a result might lead you to conclude that the coin was improperly balanced because there were 10 more heads. However, you should be concerned that maybe the difference was an accident. How can you figure the chances that such a result was an accident? Statistics allow this to be done.

* See *How to Lie with Statistics,* by D. Huff, W. W. Norton and Co., Inc., 1954.

The procedure is fairly straightforward. Let us assume the hypothesis that the coin is honest. If this is true, the chance of getting a head or of getting a tail on any single flip is one out of two, or one half. The chance of getting any combination of heads and tails can be calculated from a formula figured out a long time ago. The formula shows that if 100 flips of a coin give 50 heads and 50 tails, you can be virtually certain that the coin is honest. This result means that the hypothesis that the coin was honest cannot be rejected, and consequently we accept it.

The chances of getting 55 heads and 45 tails with an honest coin are about 32 out of 100. One must recognize that in this circumstance the 55–45 split could have been an accident. Thus you cannot reject the hypothesis that the coin is honest.

Let us say that a certain coin yields 60 heads and 40 tails. The chances (more properly, the *probability*) of getting this result with a well-balanced coin would be about 4 out of 100. Since this is such an improbable event, we are led to conclude that if the coin were honest, a 60–40 split would not occur. As a result, we reject the hypothesis that the coin is properly balanced. We say that the hypothesis has been disproved. Note, however, that we have a measure of the chances that our conclusion is in error. The smaller the probability, the smaller the chance of error. The probability value at which the hypothesis is regarded as having been disproved, a value called the *level of significance*, is not rigidly fixed. It depends on the type of problem. In most meteorological problems 5 out of 100 is usually taken as the probability value at which any hypothesis is rejected.

In summary let us note that the procedure of testing the coin involved the following steps: (1) Setting up a hypothesis–statisticians call it "the null hypothesis";

(2) designing and carrying out a series of tests; (3) calculating the probability of getting the observed results in the event the hypothesis is true; and (4) rejecting the hypothesis if the probability is below the 0.05 level of significance.

In evaluating the effects of cloud seeding a similar series of steps should be followed. The hypothesis to test is that seeding has no effect. If the results of the statistical analyses lead to the rejection of this hypothesis, we are left to conclude that the seeding either increased or decreased rainfall. The decision then depends on what is physically plausible and on the rainfall observations.

Designs of Cloud-seeding Programs

The cloud-seeding programs conducted in the early 1950s were simply designed. If a group of ranchers were interested in more rainfall over their land, a battery of silver-iodide generators was set up on the ground around the property. Whenever clouds were present, some of the generators were turned on. The ones selected were those upwind from the property.

Rainfall records were taken at various places in and around the area where more rainfall was needed.

At the end of the season, let us say the months of July and August, the quantity of rainfall over the seeded area was added up. This figure was compared with the long-term average rainfall (that is, the so-called "normal rainfall") for the same months. If the rainfall during the seeding period was above normal, it was easy to be led to the conclusion that seeding was effective.

As we have noted, however, this type of reasoning can lead us astray altogether. The fact that the normal rainfall is calculated by averaging periods with more and less rainfall than normal makes the results of an experiment such as the one described of little value in testing the effects of cloud seeding.

When this fact was recognized, a new scheme for evaluating the effects of cloud seeding was evolved. Instead of working with only a single area, two were used. One was called the *control* area, the other the *target* area. The target area was the one over which the cloud seeding was to be performed. Attempts were made to select areas close to one another and having similar types of weather. When the rainfall is high on the target, it should be high on the control area, and when low on the target should be low on the control area. In the words of the statistician, it is desirable that the rainfall on the two areas be highly correlated. In an ideal situation a diagram showing the rainfall on the two areas would look like Figure 25. The points representing the rainfall during any period of time would fall on a straight line. This diagram, then, would represent conditions when no seeding was taking place. During wet years both areas would be wet; during dry years both areas would be dry.

When cloud seeding was performed, it was restricted to the target area. If the seeding were effective, it would be expected to increase the rainfall on the target area but not on the control area, and one might observe conditions such as those represented by the circle in Figure 25. If in our ideal case such a result were to occur, we would be led to conclude that seeding had increased the rainfall. The fact that it was, let us say, a wet year is shown by the rainfall over the control area. If there were no seeding, we would have expected the quantity of rainfall over the target area to be indicated by point N. Since during the seeding year the amount over the target was that indicated by point S, the diagram would show that seeding caused the increase given by the difference between S and N.

This scheme of cloud-seeding evaluation appears to be foolproof. It has been used extensively. However, it is not nearly so good as it looks and can be misleading.

The biggest difficulty arising in this procedure is that no two areas are perfectly correlated as far as rainfall is concerned. A more realistic plot of rainfall on target

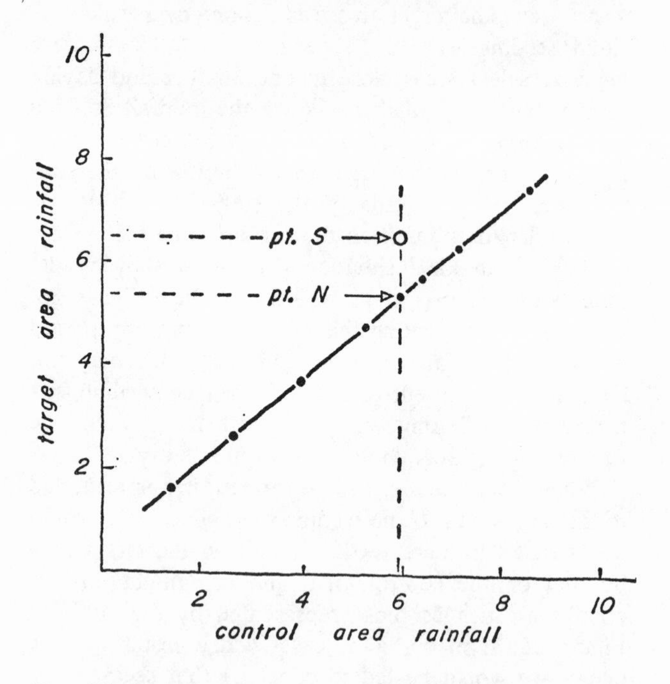

FIG. 25. A plot of the quantity of rainfall over a control and a target area in an ideal situation. Each point could represent the rainfall during any suitable time period—for example, a day, a month, or a year. The target area is the one over which cloud seeding is planned. If during the seeding period the quantities of rain over the control and target areas were represented by the circle marked S, it would be concluded the seeding had increased the rainfall.

and control areas is shown in Figure 26. The line drawn through the points is intended to represent the best estimate of the relation of target to control rainfall. One can think of it as giving average conditions. It is important, however, to realize that the points scatter around the line.

Let us assume that cloud seeding was performed in

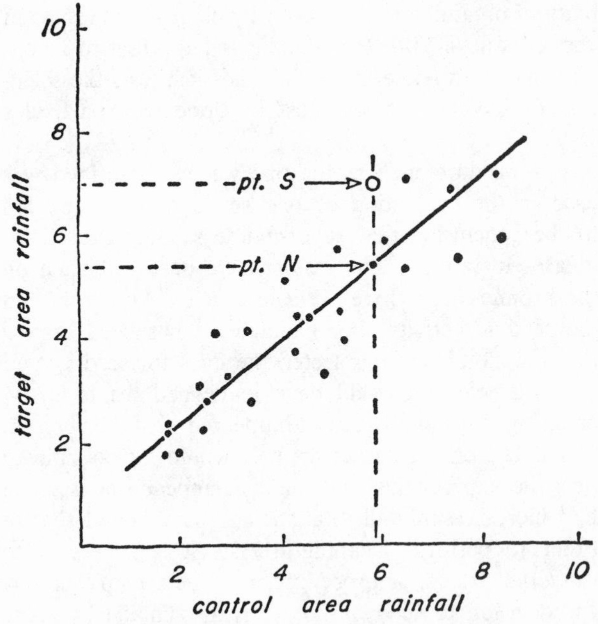

FIG. 26. A realistic plot of the rainfall over a target and a control area. The line through the points, called a "regression line," is calculated to show the best relationship between the rainfall over the two areas. If during the seeded period the rainfall amounts represented by the circle labeled S were measured, no conclusion whether or not the seeding increased rainfall could be drawn until statistical tests were made to establish the likelihood of such a result by chance. This check is necessary because there are so many points both above and below the line.

a satisfactory manner, and that the point representing the rainfall during the seeding period fell at S. It is not valid to conclude that because the point falls above the line, seeding was effective. Even if the point were to fall farther away from the line than all the other points, it still cannot be inferred that seeding increased the rainfall. Before one could arrive at such a conclusion, it would be necessary to show that the probability of obtaining such a result by chance was less than about 5 out of 100. However, it is important to recognize that in order to make the statistical tests the selection of seeding periods must be done in a particular way.

A procedure such as the one we just described was used in the evaluation of one series of tests and led to the conclusion that silver-iodide seeding caused increases in rainfall. The results have been criticized on the grounds that there were insufficient "controls" on the seeding activity. For example, it has been argued that the cloud seeder selected the days for seeding and that this selection could have influenced the tests. By choosing days on which it appeared that heavy rain would fall on the target area, it would be possible to bias the experiments and make it appear that seeding had increased rainfall. Statisticians have stated that in order to perform meaningful tests it is necessary to *randomize* them. A procedure must be set up whereby the decision to seed or not to seed is made by some mechanical arrangement such as the flip of an honest coin. If the situation is regarded as favorable, the coin is flipped, heads meaning to seed, tails not to seed. On each occasion when suitable clouds develop, the same procedure is repeated.

There are many ways to randomize a seeding experiment. The aim clearly is to prevent the investigators, either consciously or subconsciously, from influencing the results.

If it is impossible to find two nearby areas whose rainfall is highly correlated, one can examine the rainfall over a single area. Individual days or groups of days can be seeded or not seeded on a random basis.

Other aspects of the design of a cloud-seeding program are important, too. The number and location of the rain gauges must be selected to obtain a representative measure of the quantity of rainfall, for instance.

Of course, it is essential that the seeding nuclei, silver iodide usually, should enter the clouds at the proper temperature levels and in the right concentrations. When seeding is done from the ground, it is often difficult to know that these conditions are fulfilled. With airplane seeding, you can have greater confidence that these requirements are met.

All the foregoing considerations must be taken into account when a project to attempt to increase rainfall artificially is planned. Unless procedures such as these are followed, it will not be possible to determine satisfactorily whether or not the seeding had any effect on rainfall at the ground.

Finally, we should realize that no cloud-seeding scheme is of the slightest possible value unless clouds of the proper types occur in sufficient numbers and temperatures. Sometimes, in the past, ice-crystal nuclei, such as silver iodide, have been used to seed clouds which did not penetrate the freezing level. Of course, such efforts were doomed to failure at the outset.

Let us examine some of the most notable rain-stimulation tests made in the past and see what the results were.

Some Experiments to Stimulate Rainfall

If you were to tabulate the many rain-stimulation research projects conducted since 1947, you would find it convenient to separate them into two groups. The first group would include those which tackled the problem head on. Clouds or storms have been seeded and networks of rain gauges used to measure how much rain or snow reached the ground. As noted in the preceding section, this approach is not as simple as it sounds.

Another approach to the problem has been to study the effects of the seeding material on individual clouds. The aim of the experiments has been to find out if it is possible to start the precipitation process in clouds which would not have rained naturally. Instead of a network of rain gauges, radar has been the chief device for determining if and when rain or snow particles form (see Plate X). In this type of study, seeding is usually done from an airplane so that particular types of clouds can be located and seeded.

It has been more or less assumed that if you could start the rain process, a significant increase of rainfall at the ground would follow automatically. There is some question of the validity of this assumption. Raindrops formed in a cloud may partially or totally evaporate before they reach the ground. Nevertheless, from the point of view of understanding how nature makes rain, it is important to know if the addition of dry ice, silver iodide, salt, or waterdrops can cause raindrops to form in clouds.

Let us first consider some of the experiments which were concerned with rainfall initiation. We then shall come to those experiments dealing with the more practical problem, rainfall at the ground.

Starting the Rainfall Process

A group of Australian scientists, working under the leadership of E. G. Bowen, carried out extensive tests during the period 1947 to 1951. They dropped dry ice into building cumulus clouds and observed them with an airborne radar set. They reached the conclusion that it was possible to start precipitation. Similar conclusions were drawn by various other groups, including the one at General Electric led by Langmuir and Schaefer. On the other hand, in a project carried out during 1948 and 1949, Richard D. Coons and Ross Gunn, of the U. S. Weather Bureau, reported that they usually could not initiate precipitation unless other clouds within thirty miles were already raining. The implication of this result was that rainfall could not be caused to form unless conditions were just about ripe for natural rainfall. These investigators observed that the quantity of rain from the seeded clouds was very small. Their final conclusion was that cloud seeding of convective clouds could not increase rainfall by economically important amounts.

In Africa, Asia, and other parts of the world many attempts have been made to stimulate the growth of rainfall in warm clouds by seeding them with giant condensation nuclei. In some cases salt particles were shot into clouds in rockets; in others the salt was dropped from airplanes. As in the case with dry ice, some reports were positive and others negative.

Controlled Experiments in Cloud Seeding

One of the difficulties in evaluating most of these projects has been the impossibility of making satisfactory statistical tests. An important exception was an extensive research project conducted by a group of scientists at The University of Chicago, under the leadership of Horace R. Byers and Roscoe R. Braham, Jr. The aims of the program were to test the effects of dry ice on cold cumulus and of a water spray on warm cumulus. Since this project yielded convincing results, it is worth discussing it in some detail.

The research project was carefully designed with the

assistance of two statisticians, William H. Krushal and K. Alexander Brownlee. At the start of the research it was decided to seed individual clouds and use an airborne radar set to establish whether precipitation particles were formed in the clouds. It was recognized that some of the clouds selected for study would normally develop precipitation echoes even if they were not seeded. The big problem would be to interpret properly the differences between the fraction of seeded clouds with rain echoes and the fraction of not-seeded clouds with echoes.

As already pointed out, if you wish to make a proper use of statistical tests it is essential that there be sufficient controls to prevent the investigators from biasing the results. So, randomization was brought into the plan. It was decided that clouds would be studied in pairs. A randomization scheme was used to decide which of the pair of clouds was to be seeded and which one left alone.

The idea of examining pairs of clouds selected near one another and close in time has the advantage of allowing the use of fairly simple statistical tests. However, if a pairing scheme is followed, some precautions must be taken or the entire experiment may be ruined.

The most vital precaution is that the scientist who selects the cloud for study (whom we can call the controller) must *not* know if the cloud is to be seeded. If the first cloud of a pair is seeded, then the second will not be seeded, and *vice versa.* Thus, if the controller knows what action was taken on the first cloud, he also knows whether or not the second cloud will be seeded. Let us say that the first cloud of the pair was not seeded and did not develop a radar echo. If the controller wishes to show that seeding is effective, he can select for the second (and the seeded) cloud of the pair a larger and more substantial cloud. Even if the controller is entirely honest, the knowledge of what is about to transpire may still influence him in the selection of clouds.

In The University of Chicago experiments, the following procedure was employed. A B-17 airplane with a suitable airborne radar set was used. The controller of the flight sat in a special observing position just behind the pilot. His job was to direct the flight and to select clouds for study.

After the airplane had reached the seeding altitude and all the equipment had been checked, the seeding trials began. When the controller spotted a building cumulus cloud towering to an altitude where natural precipitation might possibly occur, the airplane was flown through or near the cloud. The radar operator noted whether or not there was a radar echo from it. If there was no echo, the controller instructed the pilot to fly away from the cloud for about twenty seconds, make a 180-degree turn, and head in for a penetration of the cloud. At the same time he informed the rest of the crew that a cloud had been selected for study.

In the rear compartment of the airplane another scientist was listening for instructions. This man, called the meteorological engineer, operated a battery of electronic gear for taking many types of measurements in the clouds. He also operated the seeding equipment. It is important to note that he was separated from the controller by the airplane bomb bay and the radio room. The controller could not see or hear what was taking place in the rear cabin.

When the meteorological engineer received the word that the first cloud was selected, he opened a sealed envelope. Inside was a card with the instructions "Seed the first cloud; do not seed the second," or *vice versa.* The cards were prepared by means of a suitable randomizing procedure. As the airplane made its run towards the cloud, the engineer got ready to carry out the action dictated by the card. If the card said "Seed," he stationed himself at the seeding position and waited for the final countdown.

When the airplane was about ten seconds from the cloud, the controller started counting "Ten, nine, eight. . . ." At zero, the meteorological engineer threw the switch and the cloud was seeded. The controller, up front, did not know whether the critical switch was thrown or not. The airplane then was directed through the cloud on successive passes until the cloud either dissipated or developed into an active shower. The scope of the radar set was photographed continuously for a permanent record to establish whether a precipitation echo developed.

After the first cloud had been investigated thoroughly, the controller looked for another cloud like it. The same procedure was followed, except that the meteorological engineer took the contrary action. If he seeded the first cloud, he did not seed the second.

Let us note again that the controller did not know which cloud of the pair was seeded. For this reason he would try to select pairs of clouds as much alike as possible. But even if they differed, as they almost always did to a certain extent, the randomization scheme assured the scientists that after a long series of tests the groups of seeded and not-seeded clouds would be statistically the same.

Cold cumulus clouds over the central and southwestern United States were seeded with dry ice. The number of radar echoes in seeded clouds was greater than the number in the not-seeded clouds, but the difference was small. The statistical tests showed that chance may have entered the picture.

On the other hand, the seeding of warm cumulus clouds was found to be effective. Clouds over the ocean in the vicinity of Puerto Rico were seeded with water droplets, most of which had diameters of a few hundred microns. They were produced by dumping about

400 gallons of water from the bomb bay of a B-17 during the roughly 20-second interval required to fly through the cloud. There were almost twice as many rain echoes in the seeded clouds as in the not-seeded clouds. The odds that the observed results occurred by chance were only 2 to 100. Also, it was found that in the seeded clouds the rain echoes developed in about half the time taken for the formation of echoes in the not-seeded clouds.

This series of experiments clearly showed that water-spray seeding could start precipitation in tropical cumuli. It represented an important step forward. However, it did not answer the question, can cloud seeding increase rainfall at the ground?

Increasing Rainfall at the Ground

A few articles in scientific literature and many reports of cloud-seeding companies have claimed that silver-iodide particles dispensed from ground-based silver-iodide generators caused increases of rain or snow. Nevertheless, the scientific community in the early 1950s was still uncertain whether the claims could be accepted, because many details of the work were unknown.

In 1953, the President of the United States appointed a board called the Advisory Committee on Weather Control. The Committee was established for the purpose of investigating the status of weather modification. The late Howard T. Orville was named chairman. Frederick A. Berry was chief scientific adviser, and Herbert C. S. Thom was chief climatologist. For three years the statisticians and meteorologists employed by the Committee analyzed all the data they could obtain. For the most part, it was supplied by commercial cloud seeders all over the United States. The final report of the Committee, issued by the U. S.

Government Printing Office in 1957, stated the following:

> On the basis of its statistical evaluation of wintertime cloud seeding using silver iodide as the seeding agent, the committee concluded that:
>
> (1) The statistical procedures employed indicated that the seeding of winter-type storm clouds in mountainous areas in western United States produced an average increase in precipitation of 10 to 15 per cent from seeded storms, with heavy odds that this increase was not the result of natural variations in the amount of rainfall.
>
> (2) In nonmountainous areas, the same statistical procedures did not detect any increase in precipitation that could be attributed to cloud seeding. This does not mean that effects may not have been produced. The greater variability of rainfall patterns in nonmountainous areas made the techniques less sensitive for picking up small changes which might have occurred there than when applied to the mountainous regions.
>
> (3) No evidence was found in the evaluation of any project which was intended to increase precipitation that cloud seeding had produced a detectable negative effect on precipitation.

When these conclusions were released, they produced mixed reactions. Some meteorologists regarded them as the first concrete evidence that in at least certain situations—namely, winter storms in mountainous areas—it is possible to increase rainfall by a small but very important amount. On the other hand, the conclusions were criticized by a group of statisticians who maintained that the results could not be accepted. The main point they made was that insufficient controls had been built into the experiments. In particular, they stated that the tests were not randomized and that as a result the statistical procedures employed were not valid. This position won the support of many scientists involved in cloud-seeding tests.

By 1957 it became apparent to many meteorologists in the United States that it would be desirable to carry out some rain-making tests with the necessary controls. On the west coast of the United States a joint project, which recruited the talents of a commercial cloud-seeding firm and a group of university statisticians, was started. Winter storms of the kind that appeared suitable were seeded by means of ground-based silver-iodide generators according to an acceptable randomization procedure. This program was carried on for a period of four years. For various reasons, most of them beyond the control of the investigators, the results of this project were inconclusive.

At about the same time a randomized program of airborne silver-iodide seeding was being carried out at The University of Arizona. Summer cumulus clouds over a fairly isolated mountain range were studied. After four summers of experimentation, it was still not possible to reject the hypothesis that seeding had no effect. The evidence collected from radar, camera, and rain-gauge observations suggested that the seeding may have initiated precipitation echoes without at the same time causing a measurable effect on rainfall.

In 1955 a group of Australian scientists working with E. G. Bowen started a program of airborne silver-iodide seeding. Early results of these randomized experiments were very encouraging. As the experiments continued, however, the differences of the precipitation in seeded and not-seeded periods decreased. By 1961 Bowen and his associates had reached the tentative conclusion that seeding could increase rainfall only in very special weather situations—those in which most of the clouds were cumuliform rather than stratified. This excellent work by the Australians is still in progress.

The small measure of positive results has discouraged many people who looked optimistically to cloud seeding as a means of increasing the rapidly diminishing water supplies in many parts of the world. Still, one cannot reject the idea that in certain circumstances it may be possible to increase the rainfall by the introduction of particular types of nuclei.

Preliminary results of randomized silver-iodide experiments in Switzerland, Japan, and Mexico have been encouraging. Before jumping to conclusions, we must await the completion of a fairly long series of tests in order that the effects of chance can be taken into account properly.

Several theoretical studies have led to the conclusion that particular cloud types may be amenable to treatment. Helmut K. Weickmann of the U. S. Signal Corps has stated, "Modest increases in precipitation may be achieved through seeding of warm-front precipitation." He was referring to the seeding of clouds of the altostratus and nimbostratus types, mentioned in Chapter 4.

The English scientist Frank H. Ludlam made an evaluation of the effects one might expect by seeding orographic clouds formed over a Swedish mountain range. These clouds normally are composed of supercooled water droplets which form along one side of the cloud and evaporate on the other side (Chapter 4). Much of the water contained in the form of cloud droplets is lost by evaporation. Ludlam's calculations led him to conclude that silver-iodide seeding might cause a large increase in the amount of snowfall.

Neither Weickmann's nor Ludlam's ideas have been adequately tested. Until these possibilities, as well as others having promise, have been examined, the possibility that cloud seeding may increase rainfall cannot be discarded.

However, it is of the greatest importance to recognize that much, possibly most, of the uncertainty in assessing the efficacy of cloud seeding has resulted because we still do not know enough about natural cloud and precipitation processes. When cloud-seeding experiments are conducted, they should be carried out in a way to give more information than merely whether or not the seeding was effective. They should be regarded as atmospheric experiments designed to shed some light on the physics of the problems involved.

In summary we can cite two conclusions taken from a report by the World Meteorological Organization, issued in 1955 but still applicable seven years later:

> . . . there is some justification for supposing that where certain special kinds of clouds frequently occur, skillfully conducted seeding operations might result in a local net increase in precipitation of economic benefit. . . .
>
> In our opinion, a net increase of precipitation has not been demonstrated beyond reasonable doubt in any seeding operations yet described in the scientific literature, and it seems that at least most of the claims made in other publications and in newspapers have not had adequate foundation.

Chapter 10

MODIFICATION OF HAILSTORMS

In Chapter 7 we discussed the formation of hail and mentioned the reasons why hail is an enemy of the farmer, the aviator, and just about everyone else. Scientists have been working to develop a better understanding of how and why hail is produced in some thunderstorms but not in others. They have been particularly interested in the way that large hailstones are grown. At the present time major efforts also are being devoted to finding practical techniques for suppressing the formation of damaging hailstones. Most research in this field has been aimed not so much at the total elimination of hail, but rather toward finding ways to produce a greater number of small stones instead of fewer, larger, and more damaging ones.

Use of Rockets

Various modification techniques have been employed. Since the late 1940s, farmers in the northern part of Italy have used rockets to fight hail. The rockets are about three inches in diameter and five feet long and are relatively inexpensive. They are constructed mainly of cardboard and are capable of reaching heights of about 1500 meters. At this altitude the nose cone, composed of about 800 grams of gunpowder, explodes. During the 1959 hail season more than 100,000 rockets were fired. The common practice is to start firing when a thunderstorm moves overhead and continue until the storm has passed. Rockets have also been used to fight hail in France, Switzerland, Russia, and Germany.

It is not clear how the idea of using rockets for this purpose was started. More importantly, it is not even clear how the rocket is supposed to influence the hailstones. Nevertheless, the practice has now been established.

Ottavio Vittori, an Italian scientist, has questioned many farmers about their views on this subject. He found that quite a few had been using rockets for many years and were convinced that they were effective. It was commonly reported that minutes after the shots were fired, the hailstones in the vicinity of the rocket-launching stand were found to be "mushy" rather than hard ice particles. Vittori speculated that perhaps the explosion of the rocket head produced a pressure wave which was capable of causing the formation of many tiny cracks in the ice. In this condition a hailstone should be less destructive when it hits trees and other plants. It is known that air and water bubbles are trapped inside hailstones. A sufficiently strong pressure wave might cause the effects proposed by Vittori. His ideas have been questioned by some scientists, but because of the importance of this matter the investigation continues.

Seeding with Giant Salt Nuclei

Rockets are also being used as vehicles for transporting silver iodide and salt into thunderstorms likely to produce hail. The idea of using these reagents has some scientific foundation. In Chapter 7 we noted that for the growth of large hailstones it was necessary to have an abundant supply of supercooled water. If the supply of water were limited, it would be reasonable to expect only a limited number of large stones.

It was argued that by increasing the number of hailstones it would be possible to limit their growth. The greater the number of stones sharing the available water, the less water each one would get.

There is evidence suggesting that large cloud droplets which freeze are often the embryos on which hailstones grow. If the number of such large droplets were increased, it might be possible to increase the number of stones. This is the basis on which salt seeding was founded. In theory one can increase the number of large cloud droplets by seeding the clouds with giant salt nuclei.

Some attempts have been made to make use of this idea, but the results have been inconclusive. In 1960, Frank H. Ludlam, who at one time had suggested that the scheme of salt seeding might work, rejected this viewpoint. On the basis of detailed studies of hailstorms, he reached the conclusion that the idea that there is only a limited quantity of water is not strictly correct. His new theory proposes that in order for a large stone to form it must make several trips through the cloud (Chapter 7). Almost any ice particle that manages to follow such a route is likely to encounter enough supercooled water droplets to grow to a large size.

Seeding with Silver Iodide

Another scheme for suppressing the formation of damaging hail has been based, not on the increase of the number of hail embryos, but on reducing the rate at which the embryos can accumulate ice. As we know, a hailstone grows by colliding with supercooled water droplets. It also appears that some ice crystals are captured and incorporated into the ice. However, the rate of capture of ice crystals is thought to be small. If the number of supercooled droplets in a cloud were reduced, the rate of growth of the stones should be decreased. In theory, if all the supercooled droplets were eliminated, the formation of hailstones could be prevented.

One way to decrease the number of supercooled water droplets is to seed with silver iodide in order to convert the droplets to ice crystals. The more silver-iodide nuclei dispersed, the more complete should be the conversion.

In Russia, Switzerland, the United States, and other countries, hail-forming clouds have been seeded with silver iodide. The results, so far, have not been encouraging. Conclusion (4) of the 1957 report of the President's Advisory Committee on Weather Control (see page 119) was

> Available hail-frequency data were completely inadequate for evaluation purposes, and no conclusion as to the effectiveness of hail-suppression projects could be reached.

Field tests of silver-iodide seeding conducted after 1957 also have failed to demonstrate that there is any effect.

One reason cited for the negative or inconclusive results has been that there were not enough nuclei. In a hail cloud 3 km in diameter, the number of cloud droplets between the altitudes of 5 km and 8 km, where the temperatures would be about –5°C and about –20°C, could be about 2×10^{18} droplets. This is on the small side, because in all likelihood the cloud diameter will be greater than 3 km and the number of droplets greater than the $100/cm^3$ assumed in making the calculations.

If the updraft rate is 5 m/sec, the rate at which droplets would be moving through the volume would be about 2×10^{17} per minute. Let us assume that we have a generator producing 10^{13} nuclei per gram of silver iodide. In order to supply one nucleus for each drop, it would be necessary to dispense silver iodide at a rate of about 44 pounds per minute. If the hail cloud lasted for only ten minutes, we would have to use 440 pounds of silver iodide per cloud. If we had a very efficient generator which gave 10^{14} nuclei per gram, the quantities seeded would be ten times smaller. Since it is virtually impossible to spread a given number of nuclei uniformly throughout the cloud, an even greater amount would be needed to insure that every droplet is influenced by a silver-iodide nucleus. No one has yet seeded the supercooled parts of hail clouds with such huge quantities of silver iodide. Most tests have employed 10 to 1000 times less. Thus it is not surprising that results have been inconclusive.

A seeding program employing silver iodide in the quantities that appear to be necessary would be quite expensive. Silver iodide costs about $12 per pound. Other costs involved in the seeding would at least double the expense. It appears that this scheme, even if it worked, would be prohibitively expensive. However, for learning more about hailstorms, it would be worthwhile to carry out some seeding experiments in which massive doses of silver iodide are seeded into thunderstorms likely to produce hail.

In summary, we can say that at the present time there is no known suitable means for preventing the formation of damaging hail. More research is needed to develop a better understanding of the hail-formation process. Such efforts are likely to lead to techniques of modification that are effective and not prohibitively expensive.

Chapter 11

CONCLUSION

Since time began, weather has played an important role in the lives of all living creatures. The first recorded attempts to learn about the atmosphere are those handed down by Aristotle more than 2000 years ago. Over the centuries people accepted rain and snow, winds and storms, and did little besides look and admire or hide and tremble.

Starting in the seventeenth century, scientists began to ask questions about the atmosphere, and then to seek the answers by systematic observations and experiments. It was observed that the weather followed certain patterns, that certain types of clouds were indicators of approaching storms. When the barometer showed falling pressure, cloudy skies and rain often came; rising pressure usually was a sign of good weather. It became evident that you could do more than just accept the weather as it came–you could *predict* what was about to come. This was a great step forward. It is not enough to say, "I understand"; it is equally important to be able to predict what will happen. Newton's law is important not because it explains why all the stones dropped in the past fell to the ground; it is important because it accurately predicts that all stones which are dropped in the future will fall to the ground. The ability to predict is the only really satisfying way to measure understanding.

Among meteorologists of the world the ability to predict not only was a test of understanding, but became an important end in itself. The practical values of accurately forecasting tomorrow's weather are vast indeed. The farmer, the builder, the aviator, the sailor, and many others are vitally concerned about the weather tomorrow, or even this afternoon, if a significant change is about to occur.

Over the last fifty years we have learned a great deal about the atmosphere, and this knowledge has been put to good use in the development of scientific methods of weather prediction. Of course, the forecasts are not as good as we should like them to be. The meteorologist knows this better than anyone else. But at the same time we must realize that they are more accurate than they were twenty years ago. Statistics showing a great reduction of casualties caused by severe storms–tornadoes and hurricanes–make this clear. Nevertheless, there is room for much improvement, and outstanding atmospheric scientists in many countries are working towards this end. The development of high-speed electronic computers is now making it possible to easily handle prediction equations which in 1945 would have been practically unmanageable. We can say that the problem of predicting the larger-scale features of the atmosphere, the formation and movement of storms, is well on its way towards the goal of improved forecasts.

Let us assume that in the next few decades it will be possible to make great improvements in our forecasting skills. Would this be satisfactory? Are we satisfied merely to coexist with the weather? Of course not! We should be able to exert some control over it.

Why have we not been able to make any deep dents in the armor which surrounds the weather? There are several good reasons. The major one is that we still do not sufficiently understand the natural laws that govern the formation of clouds, rain, storms, and other meteorological phenomena.

In this short book we have given a brief statement

of what we know about cloud physics, the foundation on which most weather-modification tests have been built. Clearly, there are many aspects of this subject which are poorly understood. But this lack of understanding is not to be wondered at when one considers the magnitude of the problems and the fact that it has been only in the last fifteen years or so that they have been getting very much attention. Over this period progress in the development of our knowledge has been very impressive. But we still have a long way to go.

The study of cloud physics requires the combined efforts of meteorologists, physicists, chemists, mathematicians, and engineers. It is not enough to be able to describe the pressure patterns and air motions leading to the formation of a cloud. We must know the physics involved in the formation of the nuclei, cloud droplets, ice crystals, hailstones, electric fields, and other aspects. We know that very small traces of certain types of chemicals may have profound effects on these things. If we are ever going to learn the details of the clouds and their composition, it is necessary to be able to make reliable measurements in and around the clouds. The latest developments of our engineering sciences must be brought to bear on this problem in increasing efforts. New and better instruments are needed. New and better techniques for transporting the instruments are necessary.

Right now we are in an age of explosive advances in the atmospheric sciences. Electronic computers, new and better radar sets, high-flying aircraft, and, finally, higher-flying weather satellites are becoming a part of the meteorologist's arsenal. This is just the beginning of a great advance towards the development of techniques of weather control. Our greatest need at present is to find and train young men and women who want to participate in this adventure. It will be a hard road because the atmosphere will not stand still. You have to be prepared to use imagination, to conceive new ways to learn its secrets. But it can be done, and those who share in the doing will be making an everlasting mark on the pages of human progress.

Appendix

UNITS OF MEASURE USED IN METEOROLOGY

The following conversion factors make it possible to change from one system of units to another. The abbreviations used in this book are given in parentheses. Note that the number 10 with a superscript means the number 1 followed by a number of zeros equal to the superscript. For example, 10^5 means 100,000, and 10^3 means 1000. When the superscript is negative, the number is 1 over 10 to the same power. For example, 10^{-5} means $1/10^5$ or $1/100{,}000$.

1. LENGTH

1 statute mile (mi)	= 5280 feet (ft)
	= 0.8684 nautical mile (naut. mi)
	= 1609.3 meters (m)
	= 1.6093 kilometers (km)
1 kilometer (km)	= 10^5 centimeters (cm)
	= 10^3 meters (m)
	= 3280.84 feet (ft)
	= 0.6214 statute mile (mi)
	= 0.5396 nautical mile (naut. mi)
1 meter (m)	= 10^6 microns
	= 10^3 millimeters (mm)

	= 10^2 centimeters (cm)
	= 3.2808 feet (ft)
	= 39.370 inches (in)
1 foot (ft)	= 12 inches (in)
	= 30.48 centimeters (cm)
	= 0.3048 meter (m)
1 centimeter (cm)	= 10^4 microns
	= 10^{-2} meter (m)
	= 0.3937 inch (in)
1 micron	= 10^{-4} centimeter (cm)
	= 0.00003937 inch (in)

2. MASS

1 gram (g)	= 0.03527 ounce (oz)
	= 0.002205 pound (lb)
	= 15.432 grains (gr)
1 kilogram (kg)	= 10^3 grams (g)
	= 2.2046 pounds (lb)
1 pound avoirdupois (lb)	= 16 ounces (oz)
	= 453.59 grams (g)
	= 0.4536 kilogram (kg)

3. VELOCITY

1 mile per hour (mph)	= 0.8684 nautical mile per hour (knot)
	= 1.4667 feet per second (ft/sec)
	= 0.4470 meter per second (m/sec)

	= 1.6093 kilometers per hour (km/hr)
	= 88 feet per minute (ft/min)
1 meter per second (m/sec)	= 2.2369 miles per hour (mph)
	= 1.9425 nautical miles per hour (knots)
	= 3.2808 feet per second (ft/sec)
	= 196.850 feet per minute (ft/min)
	= 3.6 kilometers per hour (km/hr)

4. PRESSURE

1 millibar (mb)	= 0.7501 millimeter of mercury (mm Hg)
	= 0.02953 inch of mercury (in Hg)
	= 0.01450 pound per square inch (lb/in^2)
1 inch of mercury (in Hg)	= 0.4912 pound per square inch (lb/in^2)
	= 33.864 millibars (mb)
1 standard atmosphere	= 1013.250 millibars (mb)
	= 760 millimeters of mercury (mm Hg)

= 29.9213 inches of mercury (in Hg)
= 14.6960 pounds per square inch (lb/in^2)

5. TEMPERATURE CONVERSION FORMULATE

Centigrade degrees (C)
Fahrenheit degrees (F)

$$C = 5/9 \times (F - 32)$$

$$F = \frac{9 \times C}{5} + 32$$

Examples: (1) Temperature = 50°F
$C = 5/9\ (50 - 32) = 5/9\ (18) =$ 10°C

(2) Temperature = –10°C
$F = 9/5\ (-10) + 32 = -18 + 32 =$ 14°F

ADDITIONAL READING MATERIAL

F. H. Ludlam and R. S. Scorer, *Cloud Study*. John Murray, Ltd., London, 1957, 80 pp.

A pictorial study of clouds. Many beautiful photographs with detailed legends describing the properties of the clouds and processes of formation.

O. G. Sutton, *The Challenge of the Atmosphere*. Harper and Brothers, New York, 1961, 227 pp.

Excellent survey of meteorology presented in terms easily understood by the nonscientist.

L. J. Battan, *The Nature of Violent Storms*. Science Study Series, Doubleday & Co., Inc., 1961, 158 pp.

An up-to-date discussion of the properties and the mechanisms of formation of thunderstorms, tornadoes, hurricanes, and cyclones.

L. J. Battan, *Radar Observes the Weather*. Science Study Series, Doubleday & Co., Inc., 1962, 158 pp.

A review of how radar is used by meteorologists for observing the weather and for research.

P. E. Viemeister, *The Lightning Book*. Doubleday & Co., Inc., New York, 1961, 316 pp.

A detailed description, written in popular style, of all aspects of lightning.

H. Neuberger, *Introduction to Physical Meteorology*. Pennsylvania State University, University Park, Pa., 1951, 271 pp.

An introductory text dealing with the physics of the atmosphere.

W. J. Humphreys, *Physics of the Air* (3d ed.). McGraw-Hill, New York, 1940.

One of the classic books in meteorology, published originally in 1920 and now in its third edition. Contains an impressive amount of detail about many aspects of atmospheric physics.

J. C. Johnson, *Physical Meteorology*. John Wiley and Sons, Inc., New York, 1954, 393 pp.

An advanced textbook covering all aspects of the physics of the atmosphere.

B. J. Mason, *The Physics of Clouds*. Oxford University Press, London, 1957, 481 pp.

A comprehensive and detailed account of modern research in the physics of clouds, including a critical appraisal of cloud-seeding experiments.

Weatherwise. American Meteorological Society, 45 Beacon Street, Boston, Mass. Six issues per year.

A journal for students and laymen. It contains short articles written by experts on all aspects of the weather. This is an excellent means of keeping abreast of the latest advances in the atmospheric sciences.

INDEX